COLLINS | Jane's

AIRCRAFT
OF
WORLD WAR II

D1506824

HarperCollins*Publishers*

Text by Jeffrey L Ethell

Thanks to Mr William Crampton of the FI for the flags
Front cover: Rex Features

HarperCollins*Publishers*
PO Box, Glasgow G4 0NB

First Published 1995
© HarperCollins*Publishers* 1995

Flags © The Flag Institute 1995

ISBN 0 00 4708490

Printed in Italy by Amadeus S.p.A.,Rome

All rights reserved. No part of this publication may be
reproduced, stored in a retrieval system, or transmitted,
in any form or by any means, electronic, mechanical,
photocopying, recording or otherwise, without prior
permission of the publishers.

CONTENTS

CONTENTS

CONTENTS

INTRODUCTION

There will never be another period in military aviation, or military history for that matter, equalling World War II for scope, numbers and impact on the world. In many ways the conflict was the pinnacle of rapid development in aviation and certainly it was the defining event of the twentieth century. An unprecedented number of aircraft were built at staggering expense. The United States, United Kingdom, Soviet Union, Germany, Japan and Italy built over 750,000 aircraft between 1939 and 1945. The United States alone built almost half that total and trained more than 1,000,000 aircrew to man them.

Governments across the globe prepared for and fought a war that accelerated aeronautical research from the biplane era of wood and fabric into the jet age. Aircraft companies used to measuring a good production run in terms of two digits were called upon to build thousands of warplanes in record time. New designs proliferated. It was a time of almost unlimited money for better machines, forcing aeronautical engineers to reach for ever higher performance in ever shorter time spans.

While most of the world slept through the Great Depression of the 1930s, battling with unemployment and isolationism, Germany and Japan built their military forces into a formidable, world-

Aircraft design advanced very quickly: ordered by the US Army in 1937, the Curtiss P-36 was obsolete by 1941

threatening presence with airpower as the tip of the spear. When the Messerschmitt 109 and Mitsubishi Zero entered service there were few aircraft like them in any nation. The United Kingdom came the closest to being prepared in the air with two outstanding fighters: the Hawker Hurricane and Supermarine Spitfire. Within the United States military aviation struggled desperately. Budgets were pitifully small, providing little equipment and even less pay for those

7

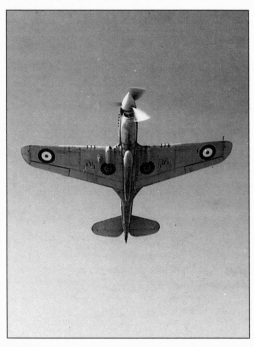

US factories turned out over 13,000 of the Curtiss P-40 series: the Kittyhawk seen here flew with the RAF

Over 100 Vickers Wellesleys were still serving with the RAF in 1939 and saw action against the Italian forces

required to operate it. Sadly, most military men had been looked upon with suspicion by a pacifist, isolationist populace with horrifying memories of World War I trench and gas warfare.

In a secret October 1924 report, US Army Air Service Brig. Gen. Billy Mitchell predicted, "the growing air power of Japan will be the decisive element in the mastery of the Pacific" through an air attack on Pearl Harbor at 7:30 AM on a Sunday with simultaneous attacks on the Philippines. Mitchell was scorned as a fanatic and a warmonger in a nation which so desperately wanted to believe World War I

In the late 1930s the RAF had 26 squadrons of Blenheim bombers capable of carrying 454 kg (1000lb) of bombs

was the war to end all wars. Even so, military planners saw what was coming and tried to prepare, often behind the scenes and without official approval.

In spite of hopes for the better, war broke out on September 1, 1939 when Germany invaded Poland. In 1940 Germany took its *Blitzkrieg* (lightning war) into the Low Countries and France. Poised to invade Britain, Hitler's victorious armies had only one obstacle to overcome, Britain's Royal Air Force (RAF). During the ensuing Battle of Britain throughout that high summer of 1940, Germany and

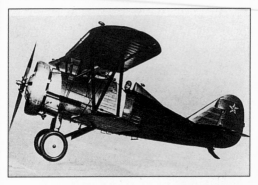

The Soviet I-5 was developed into the I-15 biplane and was still in service when the Germans invaded in 1941

Hermann Göring's vaunted *Luftwaffe* met their first defeat and the world's first battle fought entirely in the air changed warfare forever.

Even after American President Franklin D. Roosevelt called for massive increases in military aircraft and training in 1939 and 1940, most Americans wanted to believe the USA would never enter the war with Britain against Germany and that US differences with Japan could be worked out through negotiation. Aiding this apathy was an enormous lack of appreciation of the air power of the potential enemies. Just about every writer in the aviation press deprecated the quality and performance of Axis aircraft, especially those designed by the Japanese. Other media were even worse. Few westerners understood the Asiatic peoples at all, giving rise to the "Yellow Peril" caricature of the day. If familiarity breeds contempt, total lack of contact breeds total ignorance. The warnings of Charles Lindbergh and other famous flyers were universally ignored.

Entering service in 1936, the Mitsubishi A5M was the Japanese navy's first monoplane fighter aircraft

In the January 1941 issue of *Flying and Popular Aviation*, a much read American magazine, Leonard Engel reflected the prevailing attitude of the day in his article "Japan is Not an Air Power," contending, "Because the Japanese air force has devastated helpless Chinese cities does not mean it is a potent aerial armada.... Japanese carriers, surface ship for surface ship, are inferior to the British. And the British are far inferior to the American.... The stimuli of war in Europe and the armament program of the United

The Mitsubishi Ki-46 reconnaissance aircraft was faster than most Allied fighters in 1940

States have boosted air performance in the west way beyond levels attainable in Japan.... The experience of the Japanese people with mechanical gadgets is definitely limited. They have not yet gotten much beyond merely imitating what others have done. At that they are the world's finest, but imitativeness is little help in aeronautics.... The Japanese air force can be described as the sixth in the world in numbers and quality, as adequate to the job it has so far had to do. But it would not be adequate in the event of an encounter with either possible major opponent, the USA or the USSR, unless – in the case of the USSR – simultaneous war in the west drained too many planes from Siberia."

In spite of the 1940 reports from expatriate American flyer Claire Chennault in China and assistant US naval attaché Stephen Jurika in Tokyo, the American military establishment could not bring themselves to believe Japan was anything but a paper threat. As Jurika remembered,

"I had climbed into the cockpit of a Zero on display before an air show put on by the Imperial Japanese Army down at what is now Haneda International Airport.... The manufacturer's information tag in the cockpit...gave the horsepower of the engine and the weight of the aircraft. I was unable to copy down these things, but I committed them to memory

US factories built over 18,000 B-24 Liberators: a production rate unequaled by any other nation

By 1939 the Ju-86 was obsolete as a bomber, but it was developed into a high altitude reconnaissance aircraft

15

immediately and every other thing I could see in the cockpit and around the aircraft – the wing [skin]; the type of metal that was used, a very, very light aluminum; the retractable landing gear, [etc.]. I put these together and sent them [to Washington]. About three months later I received a note from someone in the Office of Naval Intelligence to the effect that I should be more careful in reporting the characteristics and estimated weight of Japanese aircraft because the

Douglas DB-7s, known to the RAF as Boston IIIs served mainly in North Africa and the Mediterranean

The Avro Manchester was handicapped by unreliable engines and only 202 were built, serving until 1943

weight was so much less than anything we had in a fighter, [while] the top speed and the g limits [were greater]. We had nothing in our own services to compare with it. Therefore, it was thought, or believed back there, that my report on the Zero was incorrect, not according to the facts.

"When I came back from Japan to the Office of Naval Intelligence, I talked to them...and the chief of naval operations [about] Japanese naval aviators and their capabilities. When I submitted myself to questions, most...were, 'don't they all have astigmatism, don't they all wear bifocals, aren't they all too short to fly, can they really do acrobatics? Oh, hell, they can't possibly be as good as we are, or half as good.' The naval aviators, the people I knew in the carrier navy, and especially the gang that was sitting there, not in ONI but in the aviation section in CNO, just would not take seriously the fact that Japanese carriers were a major threat, at least as good as ours.

"As to [Japanese] leadership, I had little to offer other than the fact that they were operating all the time. They didn't worry about 'you can only go out on Monday and come back on Friday.' If they wanted to go out and operate for six weeks at a time, they would. I think the group of aviators they had in the carrier force right at the beginning of the war was a superb group of people, really trained."

Training for battle

RAF planners like Air Marshal Hugh Dowding, the architect of victory during the Battle of Britain, realized they would need not only aircraft but the men to fly them. Creating the Commonwealth Training Plan, the RAF stretched across the seas to not only those nations under its dominion like Canada, New Zealand, Australia, Rhodesia and South

The Seafire LIII was a carrierborne version of the Spitfire which served aboard RN aircraft carriers from 1943

Africa, but to the United States. Dozens of flight training schools were created and staffed far from the dangers of combat always present over Britain, resulting in a steady flow of well-trained aircrew. Soon squadrons were formed made up entirely of Commonwealth nationals, with national shoulder patches on their RAF uniforms designating their country.

In spite of the raging conflict in Europe, the Japanese naval air attack on Pearl Harbor on 7

Tropicalised Spitfire Vcs of the RAF's 352 (Yugoslav) squadron, seen in Italy during 1944

December 1941 still caught America without enough aircraft to fight the Axis powers. Even those machines available were, on the whole, obsolete. Yet, by 1943, US industrial might was harnessed. The mighty air fleets of 1944 and 1945 launched from both land and sea, were the largest ever assembled. Never again will nations fight air battles the size of those in World War II. By the time the Japanese surrendered, almost 300,000 American aircraft had been built during the war - more than double the British production run, almost triple that of Germany and nearly five times the number pushed out in Japan. In March 1944 alone American workers built 9,113 aircraft.

Of the total American aircraft production run, almost 70,000 were built for the navy. In December 1941 the US Navy had eight battleships, four aircraft carriers, one escort carrier and 486,266 personnel. By September 1945 the numbers had swelled to 5,788 warships, 66,000 landing vessels, 16 Essex-class carriers, over 100 other carriers launched and 4,000,000 personnel, of which 400,000 (10 per cent) served the air arm. The pre-war Army Air Corps with 20,196 in uniform (11 per cent of the army) on 30

A Focke-Wulf Fw 190A-4/U-8 with wing-mounted 20 mm cannon, a 500 kg (1,102 lb) bomb and drop tanks

June 1938 grew into the massive US Army Air Force (USAAF), which six years later was swarming across the world with 2,372,292 people (31 per cent of the army).

The American training programme to get aircrew into aircraft matched industry with a frenzy: the US went from training 11,000 pilots in 1941 to training 82,700 in 1943, fifteen times the number trained in Japan during the same time period. In 1939 the navy and Marine Corps had 2,100 aircraft flown by 1,800 pilots supported by 625 ground officers and 21,000 enlisted men. By 1945 the navy had trained just short

Seen here being loaded with torpedoes, the Junkers Ju-188 was an improved version of the Ju-88 bomber

The Messerschmitt Me 210 was developed into the very capable Me 410 bomber destroyer

of 65,000 Naval Aviators, the AAF had put over 190,000 pilots and 400,000 aircrew in their aircraft, while women and minorities were becoming an integral part of the military machine.

Fighter design

Although very much a fighter campaign, air operations during World War I stamped the air commanders who were there – particularly the British and the Americans – with the indelible mark of the bomber. The ensuing two decades saw the rise of the strategic bomber which, according to its advocates,

This Yak-3 carries the tricolour of Groupe Normandie, a squadron of French volunteers in the Soviet air force

would fly so high and fast no intercepting fighter or anti-aircraft battery could shoot it down. Should some interceptor get near, every bomber would have enough guns aboard to destroy any attacker. This seemingly rendered the fighter superfluous. Despite an apparent sealed fate, fighters survived, if for no other reason than to pester enemy ground troops or intercept enemy attack aircraft.

The advocates of the bomber gained quick control of the Royal Air Force and the US Army Air Corps so that, from the mid 1920s, if an air officer wanted to

The highly manoeuverable Macchi Folgore was probably the best Italian fighter aircraft of the war

advance up the chain of command, he was compelled to tow the party line and love the bomber. There were always the odd thorns in the flesh who argued a few well-employed fighters could render a bomber formation ineffective. These unbelievers were repeatedly denounced until they either left the service or kept their peace. Nevertheless, many stayed in the service and worked behind the scenes.

With little funding or encouragement for innovation, American fighter projects officer Benjamin S. Kelsey 'invented' the term interceptor –

at least in the Army Air Corps – in order, as he recalled, "to permit development of true combat types with adequate armament and excluding baggage compartments for golf bags." Believe it or not, the latter was then an official requirement for Army fighters. Kelsey lived through those years when the fighter "was scorned by the military philosophers and neglected in the USA. In Germany and in Britain there was a more pragmatic approach which led to superior performing, record breaking Messerschmitts, Hurricanes and Spitfires."

The North American Mustang was originally ordered by the RAF: this one was lost to ground fire in 1944

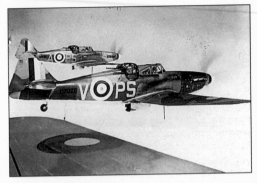

Carrying its armament in a turret, the Boulton-Paul Defiant was an unsuccessful experiment

In many ways it was a miracle the great American fighters of World War II happened at all. In the mid 1930s everything had to appear at once in the midst of the Depression with virtually no funding. Almost overnight the wire-braced, wood-structure biplane disappeared in favour of the metal monocoque, cantilever monoplane. Two 7.7 mm (0.30 in) machine guns were replaced by up to eight 12.7 mm (0.50 in) machine guns and cannons (automatic weapons exceeding 15 mm (0.59 in)). Wing and power loadings improved through aerodynamic and

27

Soviet airpower was devoted to supporting the Red Army, the Pe-2 was one of several superb light bombers developed in the USSR

propulsion innovations, making wing loadings and useful loads go up. Range went from a few hundred miles to an extreme ferry range (in the P-38) of 3,000 miles (4,827 km)

Kelsey marvelled that, with only a few exceptions, "the planes which evolved without any rational thesis still had wing areas of 225 to 250 square feet and carried one man, the pilot. Manoeuvrability, which was not defined, nevertheless dictated handy, responsive craft with good acceleration, and with adequate structural strength to withstand almost any kind of punishment which a pilot could impose."

Even so, "economy measures had so restricted engine development that all design projections of fighters during the period showed intolerable performance deficiencies. Projections of the performance requirements for the traditional small fighter dictated power plants of 1,500 hp but most nations were struggling to get over the 1,000 hp hurdle. The P-38 became the first US fighter with a real 20,000 foot critical altitude, an honest 400 mph-plus speed (644 km/h) and a realistic, hard-hitting armament of four .50s and a 37-mm (later 20-mm) cannon. The Lightning was in no way evolutionary. It was purely revolutionary, forecasting the future with metal-covered control surfaces, a nosewheel, a completely smooth stressed skin, counter-rotating propellers, a high weight-to-flat plate drag

The Messerschmitt Bf 109 equipped most German fighter squadrons in the summer of 1940

relationship, which gave it (sometimes unwelcome) fantastic dive characteristics. Its normal high wing loading of 40 pounds-plus per square foot nearly doubled existing standards.

Battle of Britain

During the Battle of Britain, British fighters proved that unescorted bombers were hopelessly vulnerable. The *Luftwaffe*, with its excellent Messerschmitt 109E, did not get the point. Instead of allowing his fighter pilots to hit the rising Spitfires and Hurricanes before they got anywhere near the bombers, Göring ordered them to stay close to their charges, thus losing the initiative. Even though several American observers, including Col. Tooey Spaatz and Kelsey, were there, most saw no need for escort fighters for American bomber groups since the B-17 Flying Fortress would be able to mount an effective self-defence. They could not have been more wrong, as they found out just three years later.

As Kelsey watched the Battle of Britain unfold, he "was present when a British commander was asked if fighters were necessary. He replied, 'Rather.' When asked if the bombers could not provide their own mutually supporting defence, he said, 'I suppose they can, but you see, when we send out the bombers without the fighters we lose the bombers. When we

The Soviet Pe-8 bombed targets behind the German lines but rarely ventured over Germany before 1945

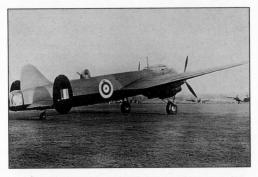

The Avro Manchester formed the basis for the Lancaster

The most famous British bomber of the war, the magnificent Avro Lancaster carried a formidable bombload and could hit targets deep into Europe

send the fighters we get the bombers back.
I don't know what the fighters do but we send them
out because we have to.

Italian Gen. Giuilo Douhet's seminal work, *The Command of the Air*, prophesied the future ability of air power to strike deep behind front lines, so crippling the enemy's war machine that surrender could be brought about without intervention of a

land army. Led by Assistant Chief of the US Army Air Service Brig. Gen. Billy Mitchell (originally in favour of smaller attack aircraft) and Marshal of the RAF Lord Hugh Trenchard, strategic bomber advocates captured the imagination of most young fliers, setting the stage for the massive bomber fleets of World War II.

The bombers of the 1920s and '30s were poor platforms from which to prove strategic bombing. As a result, advocates had very little to rely on other than strong faith and persuasive eloquence. With the Great Depression, military budgets were cut mercilessly until some planners began to wonder if there would be enough of an air force to fight the next war. Nevertheless, the Americans and the British managed to squeeze into the modern era with a few advanced aircraft, particularly the Martin B-10.

With the Boeing Model 299, entered in the Army Air Corps bomber trials of 1935, came the weapon that future luminaries Hap Arnold, Tooey Spaatz, Ira Eaker and their friends pinned their hopes on. In

A De Haviland Mosquito FV VI of 613 Squadron, lost on an intruder mission during August 1944

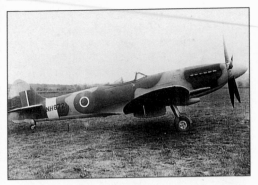

A Spitfire XIV was the first Allied aircraft to shoot down a Messerschmitt Me 262 jet fighter

spite of losing the prototype in a crash, the Army ordered a service test quantity of 13 near duplicates in January 1936 as YB-17s, already registered by Boeing as the Flying Fortress. The famous nickname seemed natural because of the heavy (for the time) defensive armament built into the design. Ironically enough, since offensive weapons were out of fashion, the long-range bomber was ordered as a coastal defence weapon, able to hit ships far out to sea. Through several long-distance flights and ship interceptions, army pilots proved their B-17s could get to targets far

35

beyond the reach of any aircraft in the world.

In spite of bitter appropriations fights on the floors of the US Congress, the Air Corps got enough money to slowly build its heavy bombardment groups and order yet another four-engine monster, the Consolidated XB-24 Liberator, which entered service in mid-1941. The new Army Air Forces, (which succeeded the Air Corps on 20 June 1941, under the command of Maj. Gen. Hap Arnold) used these types as the basis of their plans to hit Germany and Japan in the event of American involvement in the war.

Britain had learned the hard lessons of bomber operations when the RAF sent out unescorted

The Hawker Tempest II was powered by a 2,520 hp Bristol Centaurus engine

bombers to hit targets on the Continent, and during the Battle of Britain when RAF fighters attacked *Luftwaffe* bombers. In both cases defending fighters shot down the bombers in great numbers. Instead of trying to protect their bomber formations, the RAF switched to night bombing in the hopes that darkness would provide the shield they needed. Though initially equipped with twin engine types like the Vickers-Armstrong Wellington and Handley Page Hampden, RAF Bomber Command was soon flying four engine bombers like the Short Stirling, Handley Page Halifax and Avro Lancaster. The latter two, particularly the "Lanc," quickly proved to be effective

The Mosquito B. Mk IV bomber entered service in 1942 and remained in frontline squadrons until the 1950s

37

Many bombers were developed into anti-submarine aircraft: this is a Wellington XIV of Coastal Command

heavy bombers, able to carry massive bomb loads for long distances.

When the US Eighth Army Air Force arrived in Britain in the summer of 1942, RAF leaders tried in vain to talk its commander, Brig. Gen. Ira Eaker, out of what the Americans referred to as daylight precision bombing. Eaker had been an observer of the RAF's initial efforts at fighting a modern war in the air, but, in spite of the British experience, he was convinced Fortresses and Liberators had enough defensive armament to make the daylight aerial offensive feasible without extensive fighter escort.

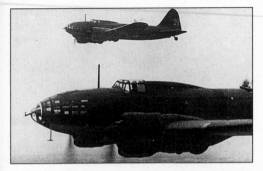

Soviet Ilyushin Il-4 bombers attacked Berlin in August 1941 but their airfields were soon overrun that autumn

This attitude was surprising for an ex-fighter pilot.

During those first months of daylight bombing, when both sides were testing each other out, Eaker was offered the new P-51 Mustang – put into production in its Merlin-powered form during April 1943 – as an escort fighter. He turned it down, still unable to see a pressing need for fighters. Both Eaker and his VIII Fighter Command leader, decorated World War I fighter pilot Brig. Gen. Frank O'D. Hunter, ignored the availability of some P-38 fighter groups in the US earlier that year, and it should not have taken half a year to discover that P-47C and early D Thunderbolts had no escort range. Eaker's stubbornness, coupled with Hunter's inertia, did not serve the Allies well. The US Army Air Force should have fielded at least a dozen groups of night bombers to get the American offensive on track.

American fighters in Britain – reverse Lend Lease Spitfires at first – were sent on ineffective Great War-style sweeps over France rather than directed to seek out the *Luftwaffe*. Even though their range would not allow much more than shallow penetrations of enemy airspace, fighters were not thought of in terms of aggressive confrontation. The policy was left intact when the first P-47s started flying combat in April 1943. This same misuse of airpower was mirrored during the North African campaign when P-38s, transferred from Britain, were sent out piecemeal on vague missions only to be slaughtered by eager German fighter pilots. This made a deep impression on Twelfth Air Force commander Brig. Gen. Jimmy Doolittle.

Far away in the Pacific Theatre another former World War I fighter pilot, US Fifth Army Air Force commander Lt. Gen. George C. Kenney, Gen. Douglas MacArthur's air force chief, had learned the lessons, turning his fighter pilots loose to pursue and destroy. Kenney had a problem, however. It was not lack of talent and foresight but a dearth of

An Il-4 of the Soviet Northern Fleet, equipped with a torpedo for anti-shipping strikes off Norway

The Westland Whirlwind was let down by unreliable engines and only 112 were built

41

equipment. Always on the short end of the stick, begging for every last morsel left over from the European war and Navy appropriations, Kenney turned to tactical cunning, specializing in low-level medium bomber attacks covered by as many fighters as he could scrape up.

When the P-38 went into action over New Guinea in December 1942, Kenney had found the weapon he was looking for and he proceeded to dominate the Southwest Pacific Area (SWPA) with it. By June 1944 he was head of the Far East Air Forces (FEAF) and on his way to Tokyo, giving as much support to MacArthur's push as he could muster. After the war MacArthur fondly recalled, "Through [Kenney's] extraordinary capacity to improvise and improve, he took a substandard force and welded it into a weapon so deadly as to take command of the air whenever it engaged the enemy."

By mid 1943 Eaker and Hunter were watching the *Luftwaffe* tear the Eighth Air Force apart until the notorious Schweinfurt mission of 14 October 1943 spelled out the obvious. The daylight offensive would have to be halted unless fighter cover could be maintained all the way to the target and back. Bombers proved to be hopelessly vulnerable to fighters and flak. Those P-38s wasted in North Africa could have been protecting the bombers from mid

The Focke-Wulf Fw 190F series was a specialist ground attack version with four bomb racks

1942, long before the first P-47s could pose a serious escort threat. As it was, Lightnings did not get back into the theatre until October 1943. Under great pressure from Arnold, furious over the mismanagement his old friend was displaying, Eaker finally began to wake up, transferring Hunter out and replacing him with Maj. Gen. William E. Kepner.

Unlike his predecessor, Bill Kepner had learned to apply the lessons from aviation in World War I, even though he had been an infantry officer in the conflict. In spite of being hampered with orders to keep his fighters tightly connected to the bombers, Kepner, remembered as 'a little bantam rooster, a gung-ho!, hell-for-leather type', encouraged the aggressive nature of his pilots, turning them loose not only to hunt

The B-24 was developed into the PB4Y Privateer long range maritime patrol aircraft

down the *Luftwaffe* in the air but on the ground. When the first Merlin-powered P-51B Mustangs arrived in November 1943, assigned to the Ninth Air Force for the forthcoming invasion, Kepner quickly managed to have them detached to the Eighth for bomber escort duty. The tide finally turned in

December 1943 when Arnold initiated action to relieve Eaker, transferring him to the Mediterranean while moving Lt. Gen. Jimmy Doolittle to command of the Eighth Air Force.

When Doolittle arrived in January he paid a visit to Kepner and noticed a lingering sign over the office

Vickers Wellingtons attacked Germany in September 1939 and flew their last bombing mission in March 1945

door, saying VIII Fighter Command's job was to protect the bombers. Doolittle snapped at Kepner, ordering him to tear the sign down and replace it with one restating their assignment was to roam free of the bomber stream and destroy the *Luftwaffe*. Doolittle later recalled during a conversation with the author, "Adolf Galland said that the day we took our fighters off the bombers and put them against the German fighters – that is, went from defensive to offensive – Germany lost the air war. I made that decision and it was my most important decision during World War II. As you can imagine, the bomber crews were upset. The fighter pilots were ecstatic."

From that point on, the USAAF daylight bombing offensive was able to do its job in concert with the RAF night offensive (called Round-the-Clock Bombing), helping to bring Germany to its knees. But in spite of the increased protection, no other American military component had comparable losses. The Eighth Air Force lost 7.42 per cent of its crews to combat deaths, far overshadowing 2.94 per cent for the Marine Corps, 2.08 per cent for the Army and .88 per cent for the Navy. For the Eighth and Fifteenth Air Forces combined, that translated to

The arrival of the P-51D made it increasingly difficult for German fighters to attack American heavy bombers

47

24,288 killed, 18,699 missing in action, 18,804 wounded and 31,436 taken prisoner. Despite the cover of darkness, the crews of RAF Bomber Command were slaughtered with the same regularity. Out of a total of 105,000 aircrew, 55,000 were killed or missing, 9,800 became POWs and 8,400 were wounded. Being in a bomber over Germany, day or night, was a very dangerous way to fight a war. There were no airborne foxholes.

The air forces in the Pacific fought a rough war as well, specializing in medium attack with B-25s and A-20s and long-range strategic bombing with B-24s and B-29s. The culmination of daylight bombing theory came with the dropping of the atomic bombs on Hiroshima and Nagasaki.

Since the end of the war the debate has raged over the effectiveness of strategic bombing. Was it a waste or did it win the war? The answer lies somewhere in the middle but is best summed up by those under the bombs. Hitler's production and armaments genius, Albert Speer, remembered, "The Hamburg operation [of 1943] had catastrophic consequences for us.... The devastation could only be compared with the effects of a major earthquake. I informed Hitler that armaments production was collapsing and threw in a further warning that a series of attacks of this sort, extended to six more cities, would bring Germany's

The turbojet-powered Gloster Meteor entered service with the RAF in 1944 and served again in Korea

armaments production to a total halt."

When the Allies finally decided to make a high priority of targeting oil production, the enemy war machine became impotent. Had it not been for relentless bombing of enemy industry and cities, World War II would have lasted longer than it did. The aircraft had come of age, evolving from the wood and fabric biplane to the jet aircraft in a brief five years. The resulting stimulus to modern technology can be felt even to this day.

49

Commonwealth Boomerang

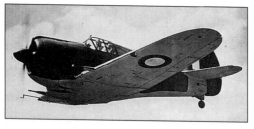

The Boomerang was the first fighter aircraft to be developed and built in Australia

The Commonwealth Aircraft Corporation was created in 1936 on the initiative of the Australian government which was painfully aware that Australia was entirely dependent on aircraft from overseas. There were grave doubts as to the ability of the British forces to defend Australia should it come to war with Japan. The Boomerang was a stopgap fighter, based on the Wirraway light combat aircraft (in turn a development of the Harvard trainer). It was not expected to equal the Japanese opposition from the very start, but its development was rapid from December 1941 until first flight in May 1942. With a 1,200 hp engine, the Boomerang did not have a very good top speed but it was reliable, rugged, manoeuvrable and easy to support in the field.

When it came up against the Zero-Sen it did well enough to keep things even: quite a feat for an aircraft obsolete from birth. Most RAAF (Royal Australian Air Force) Boomerang squadrons were sent to New Guinea where the Allies were holding the Japanese, but just barely. With no bombs, Boomerangs were used to strafe enemy positions, something they did quite effectively with two 20 mm and four .303 guns. Friendly troops came to love the aircraft and its pilots for their ability to get quite close to the front line, sorting out friend from foe. But until better aircraft were received, there was little prospect of defeating the Japanese in the air.

Commonwealth kept building Boomerangs, up to the CA-19, into early 1944. Those examples which did not see combat were used as advanced fighter transition trainers.

Specification (CA-12)

Powerplant: Pratt & Whitney R-1830-S3C4G 1,200 hp 14-cylinder radial.

Dimensions: length 7.78 m (25 ft 6 in); height 3.51 m (11 ft 6 in); wing span 11.06 m (36 ft 3 in).

Weights: empty 2,474 kg (5,450 lb); operational 3,450 kg (7,600 lb).

Performance: maximum speed 476 km/h (296 mph); service ceiling 8,845 m (29,000 ft); range 1,496 km (930 miles).

Armament: two 20 mm cannon, four .303 calibre machine guns.

Dewoitine D.520

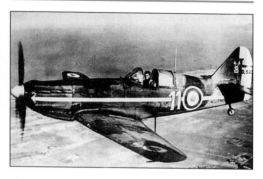

The highly manoeuvrable D.520 was handicapped by its underpowered engine

Alhough, after flying in October 1938, it was younger than the Spitfire, Hurricane and Messerschmitt 109, and a contemporary of fighters like the P-38 and the Zero-Sen, the D.520 was far inferior. Yet the small fighter was the finest produced in France during the war, more manoeuvrable than most of the opposition. The main reason for its poor performance was a shortage of the design specified 1,300 hp engine. Nevertheless, D.520s fought the Germans valiantly until France fell, gaining slightly less than a two to one kill ratio. Had French industry not suffered so badly at the hands of a socialized prewar government, far more 520s would have been available as the only near-competitor to the Bf 109 during the Battle of France. As it was, the combination of delayed orders and an antiquated production system, still dependent on skilled craftsmen rather than modern mass production methods, left the French air force with an acute shortage of first rate fighter aircraft.

After the surrender of France, the Vichy government continued production, but when the Germans confiscated over 400 of the run the aircraft were sent to Italy, Rumania and Bulgaria. After the Allied invasion in 1944 several 520s were reclaimed by Free French units and turned back against their captors in southern France. Pilots were universal in their praise of the aircraft's handling qualities, which enabled them to meet better aircraft on equal terms, but it was an example of too little too late.

Specification (D.520)

Powerplant: Hispano-Suiza 12Y-45 910 hp 12-cylinder inline.
Dimensions: length 8.76 m (28 ft 8.5 in); height 3.43 m (11 ft 3 in); wing span 10.21 m (33 ft 5.75 in).
Weights: empty 2,102 kg (4,630 lb); operational 2,802 kg (6,173 lb).
Performance: maximum speed 529 km/h (329 mph); service ceiling 11,000 m (36,090 ft); range 1,250 km (777 miles).
Armament: one 20 mm cannon, four 7.5 mm machine guns.

Morane-Saulnier M.S.406

The MS 406 was practically obsolete before it entered in service in 1939

Based on the earlier M.S.405 monoplane, the 406 entered Armée de l'Air service in March 1939, far too long after the 405's August 1935 first flight. Not only was it underpowered, typical of all French nationalized 1930s designs, but, unlike its D.520 stablemate, it did not have the handling or the armament to compensate.

Pilots had very little complimentary to say about the 406 other than it was docile and quite easy to fly with no mean streak. About the only saving grace was that there were more of them than any other French fighter when Germany invaded and most fighter squadrons were flying them. Despite their efforts, the German ground forces continued their victorious advance and France surrendered.

When the Vichy Government took over the production lines, the 406 was modified to carry external fuel and sent to the Middle East to fight the Allies. (It was feared that Vichy control of the Lebanon might be exploited by the Germans as part of their North African strategy, and British troops were ordered in during 1941.) Several other nations, including Finland, Croatia and Turkey, flew the aircraft in service but there were few sparkling reports, other than from the Finns who fitted skis and more powerful Russian engines, which improved its performance substantially. Nevertheless, the 406 never became much more than cannon fodder for skilled German fighter pilots.

Specification (M.S.406C-1)

Powerplant: Hispano-Suiza 12Y-31 860 hp 12-cylinder inline.
Dimensions: length 8.16 m (26 ft 9.25 in); height 2.84 m (9 ft 3.75 in); wing span 10.62 m (34 ft 9.75 in).
Weights: empty 1,902 kg (4,189 lb); operational 2,472 kg (5,445 lb).
Performance: maximum speed 486 km/h (302 mph); service ceiling 9,400 m (30,840 ft); range 800 km (497 miles).
Armament: one 20 mm cannon, two 7.5 mm machine guns.

Potez 63

In 1939 the French air force had 205 Potez 631 twin-engined fighters

From its first flight in April 1936, the Potez 63 series of aircraft revealed more promise than most 1930s French designs, all of which suffered under nationalized industry. In production the 631 was a fighter and the 633 a bomber. During the Battle of France several hundred 631 fighters fought the Germans, achieving a successful kill ratio in spite of a lower top speed than the Messerschmitt Bf 109.

With two 20 mm and six 7.5 mm forward firing guns, the 631 was one of the most heavily armed fighters then in service. Pilots needed but a few bursts on target to do significant damage. The aircraft was also rugged enough to withstand considerable damage, The bombers, particularly those flown by the Greeks against the Germans and the Rumanians against the Russians, were used to good effect as well. With the 63.11 came a good reconnaissance platform which was used by the Germans, the Vichy French and the Free French, all at the same time, often in the same theatres of action against each other. It was one of the most heavily-armed bombers in service at the beginning of the war, which stood it in good stead as fighters began to acquire more powerful armament.

Unlike most French wartime aircraft, the 63 series lasted through to the end due to its versatility and adaptability. Pilots found the aircraft handled well and crews depended upon its simple, rugged nature to get them through.

Specification (Potez 631)

Powerplants: two Gnome-Rhône 14M 700 hp 14-cylinder radials.
Dimensions: length 11.08 m (36 ft 4 in); height 3.60 m (11 ft 9.75 in); wing span 17.23 m (56 ft 6 in).
Weights: empty 2,601 kg (5,730 lb); operational 3,739 kg (8,235 lb).
Performance: maximum speed 439 km/h (273 mph); service ceiling 10,004 m (32,800 ft); range 1,000 km (621 miles).
Armament: two 20 mm cannon, six 7.5 mm machine guns, two flexible rear firing 7.5 mm guns.

Fokker D.XXI

A Fokker DXXI bearing the distinctive swastika markings of the Finnish air force

When it first flew in March 1936, the D.XXI, the last in a long and famous line of fighters, was already out of date with fixed landing gear and a fabric-covered steel tube rear fuselage. Nevertheless, Holland, Finland and Denmark bought the Fokker as their major fighter just before the beginning of World War II. The D.XXI, in spite of its major drawback, was an fine, rugged, manoeuvrable fighter with excellent armament for its day. As a result of its promise, there was to be a retractable gear version, which would have been put it on a par with the contemporary steel tube, fabric fuselage, wooden winged Hurricane.

Finnish pilots found that their D.XXIs did very well against the Russians after the Soviet invasion of November 1939. While Soviet air power had been undermined by Stalin's purges, the Russians were flying some very capable aircraft, so this was no mean achievement.

Dutch Fokkers fought the Germans valiantly in 1940 but were quickly worn down by superior numbers and better technology. Even so, barely 30 D.XXIs held out for several days until they ran out of ammunition and time. Dutch pilots were unanimous in their praise of the fighter, which, on paper at least, should not have done as well as it did against the redoubted Messerschmitt Bf 109. It was quite an effective bomber killer, although handicapped by its relatively light armament. Against the Ju-87 Stuka, it was very effective.

Specification (D.XXI)

Powerplant: Bristol Mercury VIII 830 hp 9-cylinder radial.
Dimensions: length 8.21 m (26 ft 11 in); height 2.95 m (9 ft 8 in); wing span 11.01 m (36 ft 1 in).
Weights: empty 1,444 kg (3,180 lb); operational 2,052 kg (4,519 lb).
Performance: maximum speed 460 km/h (286 mph); service ceiling 11,000 m (36,090 ft); range 950 km (590 miles).
Armament: four 7.9 mm machine guns.

Cant Z.1007 Alcione (Kingfisher)

Poorly armed and lightly built, the Z.1007 proved highly vulnerable to Allied fighter aircraft

A landplane bomber version of the three-engine Z.506B Airone (Heron), the all-wood Cant (for Cantieri Riuniti dell' Adriatico) Z.1007 flew for the first time in May 1937 and entered service with the Italian Regia Aeronautica in 1939. Almost half the production run had twin vertical tail surfaces. By the beginning of World War II the bomber was outdated, though its pilots lauded its excellent handling. Unfortunately, this relatively slow bomber was vulnerable to enemy fighters and it was considered an easy kill by British and American pilots who found them. The defensive armament of a mixture of four 7.7 mm and 12.7 mm machine guns in dorsal, ventral and fuselage positions was hopelessly inadequate. The only effective defence was a formation of escort fighters, in short supply most of the time.

Operating on the Mediterranean and the Russian front, the Alcione could carry over 4,000 pounds of bombs, similar to other medium bombers on both sides. The Z.1007bis had a lengthened fuselage, enlarged wings and strengthened landing gear, resulting in quite an overall improvement. The final version of the aircraft, the all metal Z.1018 Leone (Lion), had two more powerful twin engines in place of the three smaller Piaggio radials but it did not see much service before Italy surrendered in September 1943.

Specification (Z.1007bis)

Powerplants: three Piaggio P.XIbis RC40 twin-row 1,000hp 14-cylinder radials.

Dimensions: length 18.4m (60ft 4in); height 5.22m (17ft 1.5in); wing span 24.8m (81ft 4in).

Weights: empty 8,630kg (19,000lb); operational 13,620kg (30,029lb).

Performance: maximum speed 448km/h (280mph); service ceiling 8,100m (26,500ft); range 1,280km (800miles) with full bomb load.

Armament: two 12.7mm machine guns in dorsal and ventral positions, two 7.7mm machine guns in two side fuselage positions; 2,000kg (4,410lb) internal bomb load or two torpedoes and up to four 250kg (550lb) underwing bombs.

Fiat C.R.42 Falco (Falcon)

This CR.42 was shot down over Suffolk in November 1940, repaired and evaluated by the RAF

An open cockpit anachronism by the time World War II began in September 1939, the C.R. did not even enter service with the Regia Aeronautica until two months later in November. Even when initial design work on the Falco began in 1936 as a development of the C.R.41, worldwide aeronautical design was leaving the biplane behind for the all-metal monoplane. Nevertheless, a market remained for the elegant aircraft with examples eventually being sold to Belgium, Finland, Hungary and Sweden.

Pilots, even if they knew they were outclassed,

enjoyed flying the C.R.42, as long as they could stick to close air support work and stay away from enemy fighters. Most Italian Falcos were stationed in North Africa and the Mediterranean but a detachment of 50 were flown from Belgium during the Battle of Britain from October 1940 to January 1941. They were slaughtered by British fighters. Production of the Falco was finally stopped in early 1942 and the type was withdrawn from front line service as quickly as possible. From that point on most Falcos flew as lead-in fighter transition trainers for pilots on their way to combat groups and this is probably where the aircraft was most appreciated due to its sparkling handling characteristics.

Specification (C.R.42)

Powerplant: Fiat A.74 RC38 840 hp 14-cylinder radial.
Dimensions: length 8.25 m (27 ft 1.25 in); height 3.35 m (11 ft); wing span 9.7 m (31 ft 10 in).
Weights: empty 1,720 kg (3,790 lb); operational 2,300 kg (5,070 lb).
Performance: maximum speed 430 km/h (267 mph); service ceiling 10,500 m (34,450 ft); range 775 km (481 miles).
Armament: C.R.42 one 7.7 mm and one 12.7 mm machine guns atop fuselage; C.R.42bis two 12.7 mm machine guns; C.R.42ter four 12.7 mm guns (two in underwing pods); C.R.42AS two to four 12.7 mm guns and two 110 kg (220 lb) bombs on wing racks.

Fiat G.50 Freccia (Arrow)

The Fiat G.50 fought mainly in North Africa, but its light armament proved a major weakness

Developed in competition with several other designs and first flown in February 1937, the G.50 was Italy's first single-seat, all-metal monoplane fighter. By early 1938 the Freccia was in service with Regia Aeronautica fighter squadrons and several flew in the Spanish Civil War with the Aviazione Legionaria. In spite of insufficient armament, the aircraft, typical of most Italian designs, was very manoeuvrable. The primary version, the G.50bis, was improved but the real change came with mating the German Daimler-Benz DB 605A-1 engine to the airframe, creating the G.50V which became the G.55 Centauro.

The Centauro (Centaur) which flew for the first time in April 1942, was an excellent fighter with most of the modern features and speed common to World War II aircraft. By early 1943 production was underway with deliveries the following August. Fighter pilots loved the new aircraft but by the time the war ended only 105 had been built. The final version of the aircraft, with an improved 1,750 hp DB 603A engine, was designated the G.56 but only one prototype flew, in March 1944. With a top speed of 426 mph (685 km/h), it outmanoeuvred both the Me 109G and the Fw 190A in tests.

Specification

Powerplant: (G.50) Fiat A.74 RC38 840 hp 14-cylinder radial, (G.55) Daimler-Benz DB 605A 1,475 hp 12-cylinder inline.

Dimensions: length (G.50) 7.79 m (25 ft 7 in), (G.55) 9.37 m (30 ft 9 in); height (G.50) 2.9 m (9 ft 8 in), (G.55) 3.15 m (10 ft 3.75 in); wing span (G.50) 10.97 m (36 ft), (G.55) 11.85 m (38 ft 10.50 in).

Weights: empty (G.50) 1,900 kg (4,188 lb), (G.55) 2,900 kg (6,393 lb); operational (G.50) 2,706 kg (5,966 lb), (G.55) 3,710 kg (8,179 lb).

Performance: maximum speed (G.50) 471 km/h (293 mph), (G.55) 620 km/h (385 mph); service ceiling (G.50) 10,000 m (32,810 ft), (G.55) 13,000 m (42,650 ft); range (G.50) 1,000 km (621 miles), (G.55) 1,600 km (994 miles).

Armament: (G.50) two 12.7 mm machine guns, (G.55) same plus one to three 20 mm cannon.

Fiat B.R.20 Cicogna (Stork)

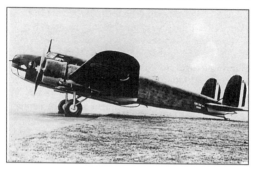

BR.20Ms took part in the latter stages of the Battle of Britain, bombing east coast ports

When it entered service in late 1936, the stressed skin B.R.20 medium bomber was quite advanced, offering the Regia Aeronautica a significant leap in capability compared to the bombers of other nations. This was not lost on potential customers, including Hungary, Japan, Spain and Venezuela who bought B.R.20s for their air forces. In the summer of 1937 the Cicogna was sent to Spain to fly with the Aviazione Legionaria, forming the backbone of Nationalist bombing operations in combination with the He 111. The success of the Fiat and the Heinkel was due in large measure to lack of effective fighter opposition, something very few appreciated. The Japanese used their B.R.20s successfully in China, and to a limited extent in the Pacific as well.

With the outbreak of World War II the B.R.20 was showing its age. Nevertheless, it was sent to Belgium in October 1940 to fly against Britain and was quickly slaughtered by British fighters due to light and ineffective defensive armament and poor fighter escort. When Germany invaded Russia several Cicogna units were sent to bolster the Luftwaffe but, once again, high losses dictated a rapid withdrawal from combat.

Relegated to maritime patrol and operational training for bomber crews, by 1942 the Fiat was put into combat in only the most favourable situations and when other types were unavailable.

Specification: (B.R.20M)

Powerplants: twin Fiat A.80 RC20 1,100 hp 18-cylinder radials.
Dimensions: length 16.78 m (55 ft); height 4.75 m (15 ft 7 in); wing span 21.56 m (70 ft 9 in).
Weights: empty 6,700 kg (14,770 lb); operational 10,450 kg (23,038 lb).
Performance: maximum speed 430 km/h (267 mph); service ceiling 6,750 m (22,145 ft); range 2,000 km (1,243 miles).
Armament: one 12.7 mm machine gun in nose turret, one 7.7 mm machine gun in ventral position, two 7.7 mm guns in dorsal turret; bomb load 2,500 kg (5,511 lb).

Macchi C.200 Saetta (Thunderbolt)

Many Italian fighter squadrons in North Africa were equipped with the tough Macchi C.200

As with Supermarine and the Spitfire, Aeronautica Macchi produced their nation's finest wartime fighters after building several outstanding Schneider Cup seaplanes. However, their first try, Mario Castoldi's (M.C.) 200, was underpowered and underarmed for a modern fighter, even long after its first flight in December 1937. After initial entry in squadron service in 1939, the Saetta's open cockpit (which, on pilot demand, replaced the original enclosed canopy) seemed a throwback but, typical of most Italian fighters, manoeuvrability was outstanding.

From the time Italy declared war against the Allies in June 1940 until her surrender in September 1943, the M.C.200 flew more operational sorties than any of the nation's aircraft. Ranging over Greece, Yugoslavia, North Africa and across the Mediterranean, then into Russia to support the Germans on the Eastern Front (where it had an excellent kill to loss ratio of 88 to 15), the Saetta could dogfight with the best Allied types and come out on top.

The Spitfire was the only Allied fighter that could outclimb the Saetta. Its very strong all-metal construction and air-cooled radial engine made the aircraft ideal for ground attack and several units flew it as a fighter-bomber. Over 1,000 were built by the time the war ended.

Specification (M.C.200)

Powerplant: Fiat A.74 RC 38 870 hp 14-cylinder radial.
Dimensions: length 8.2 m (26 ft 10.5 in); height 3.38 m (11 ft 6 in); wing span 10.58 m (34 ft 8.5 in).
Weights: empty 1,900 kg (4,188 lb); operational 2,350 kg (5,182 lb).
Performance: maximum speed 501 km/h (312 mph); service ceiling 8,900 m (29,200 ft); range 570 km (354 miles).
Armament: two 12.7 mm machine guns, with two 7.7 mm machine guns later added in wings, two 160 kg (353 lb) bombs or drop tanks.

Macchi C.202 Folgore (Lightning)

An MC.205 wearing the roundels of the Italian co-belligerent air force fightng with the Allies from 1943

Without much change to the basic airframe, the M.C.200 was turned into the M.C.202 by replacing the older, underpowered radial engine with an Alfa Romeo licence-built German DB 601A. The prototype flew in August 1940 and the Folgore went into service with the Regia Aeronautica in July 1941. Right away the aircraft proved to be among the finest fighters of the war with outstanding speed, excellent manoeuvrability (typical of all Italian fighters) and modern armament of four 12.7 mm machine guns, the equal of the American .50-calibre Browning. Later models without wing guns carried two 20 mm cannon in underwing pods. During its initial combats over the Western Desert the Folgore was quite a surprise to British pilots and it remained a respected adversary.

In April 1942 the advanced M.C.205V Veltro (Greyhound) flew with a more powerful DB 605A engine and German MG 151 20 mm cannon in place of the 7.7 mm machine guns. When Italy surrendered in September 1943, many pilots and their Macchis stayed with the Germans in the north as the Aviazione Nazionale Republicana and ended up fighting their brothers in the south who were flying Macchis with the Allied Co-Belligerent air force. The Macchis fought well until the end of the war.

Specification

Powerplant: (202) Alfa Romeo RA1000 RC41-1 (licence-built DB 601A-1) 1,175 hp 12-cylinder inline; (205) Fiat RA1050 RC58 (licence-built DB 605A-1) 1,475 hp 12-cylinder inline.
Dimensions: length 8.85 m (29 ft .5 in); height 3.04 m (9 ft 11.5 in); wing span 10.58 m (24 ft 8.5 in).
Weights: empty (202) 2,350 kg (5,181 lb), (205) 2,581 kg (5,691 lb); operational (202) 3,010 kg (6,636 lb), (205) 3,408 kg (7,514 lb).
Performance: maximum speed (202) 595 km/h (370 mph), (205) 642 km/h (399 mph); service ceiling 11,000 m (36,000 ft).
Armament: (202) two 12.7 mm and two 7.7 mm machine guns, two 160 kg (353 lb) bombs or drop tanks,; (205) same with two 20 mm cannon in place of 7.7 mm machine guns.

Reggiane Re 2000/2002 Falco (Falcon)

The highly manoeuvrable Re.2002 was closed based on the American Seversky P-35 fighter design

Almost a direct copy of the Seversky P-35, the Re 2000 flew for the first time in 1938 and when tested against the Macchi C.200 and the Bf 109E, it came out ahead on all counts. Unfortunately, it was not as rugged as the Macchi and lost the competition, forcing the company to offer it for overseas sales. Some were sold to Hungary, Sweden and Germany. The twelve 2000s under the Italian flag served with the Navy and were based in Sicily. The Re 2001 was re-engined with the Daimler-Benz DB 601A and entered service with the Regia Aeronautica over Malta in May 1942 but only 252 were built due to the shortage of engines. The Re 2002 Ariete entered combat in 1942 and flew numerous sorties against the

Allied landings in Sicily but only around 50 were built. The final version of the series was the Re 2005 Sagittario (Archer) fitted with a Fiat-built DB 605A-1 engine. It was assigned to the defence of Rome before the Italian surrender. Those that stayed with the German forces were used as interceptors in Bucharest and even Berlin due to their ability to get to 20,000 feet in just over four minutes.

Specification

Powerplant: (2000) Riaggio P.XIbis RC40 1,025 hp 14-cylinder radial; (2001) Alfa Romeo RA.1000 RC41 (DB 601) 1,175 hp 12-cylinder inline; (2002) Piaggio P.XIX RC45 1,025 hp 14-cylinder radial; (2005) Fiat RA.1050 RC58 (DB 605) 1,475 12-cylinder inline.

Dimensions: length (2000) 7.95 m (26 ft 2.5 in), (2001/2002) 8.2 m (26 ft 10 in), (2005) 9.1 m (28 ft 7.75 in); height 3.15 m (10 ft 4 in); wing span 11 m (36 ft 1 in).

Weights: empty 1,905 kg (4,200 lb); operational (2000) 2,595 kg (5,722 lb), (2001 to 2005) up to 3,650 kg (7,850 lb).

Performance: maximum speed (2000/2001/2002) 540 km/h (335 mph), (2005) 630 km/h (391 mph); service ceiling (2000/2001/2002) 11,200 m (36,745 ft), (2005) 12,250 m (40,000 ft); range (2000/2001/2002) 950 km (590 miles), (2005) 1,250 km (787 miles).

Armament: (2000) two 12.7 mm machine guns and one 200 kg (441 lb) bomb, (2001) same with two 7.7 mm machine guns or two 20 mm cannon and one 640 kg (1,410 lb) bomb, (2005) same with three 20 mm cannon.

Savoia-Marchetti S.M.79 Sparviero (Hawk)

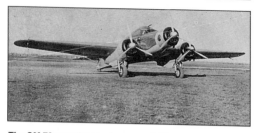

The SM.79 was the most successful Italian bomber of World War II, with over 1300 being built.

In spite of its cumbersome appearance and outdated steel tube/wood/fabric construction, the S.M.79 was a rugged, reliable multi-role medium bomber which did quite a bit of damage in the face of heavy opposition. Developed from a civil airliner, the first Sparvieros entered service with the Regia Aeronautica in late 1936, just in time to fly combat over Spain with the Aviacion Legionaria, the Italian contingent fighting in support of the Nationalists. Its performance drew favourable comments from both sides, leading to a succession of export orders

An effective torpedo bomber as well, the S.M.79 served in the air forces of Brazil, Iraq, Yugoslavia, Romania and Spain, some right up to the end of the war. The Romanians flew them on the Russian front from 1941 to 1944, an unprecedented record for an aircraft designed in the early 1930s.

Though known as a tri-motor, several versions were built as twin-engined aircraft using a number of different powerplants, including Junkers Jumo 211D 1,220 hp inlines. Regardless of the version, its handling pleased most pilots and its ability to come home with extensive damage endeared it even more. Used throughout North Africa and the Mediterranean until the Italian surrender in September 1943, the Sparviero remained flying with both the Italian co-belligerent forces fighting alongside the Allies and the surviving pro-Nazi units.

Specification (S.M.79-II)

Powerplants: three Alfa Romeo 126 RC34 780 hp 9-cylinder radials, later three Piaggio P.XI RC40 1,000 hp 14-cylinder radials.

Dimensions: length 16.2 m (53 ft 1.75 in); height 4.1 m (13 ft 5.5 in); wing span 21.2 m (69 ft 6.5 in).

Weights: empty 7,600 kg (16,755 lb); operational 11,300 kg (24,192 lb).

Performance: maximum speed 434 km/h (270 mph); service ceiling 7,000 m (23,000 ft); range 2,000 km (1,240 miles).

Armament: three 12.7 mm machine guns mounted ahead above cockpit, in dorsal position and ventral gondola; one 7.7 mm machine gun fired from either rear fuselage window; internal bomb load 1,000 kg (2,205 lb) or two external torpedoes.

Aichi D3A "Val"

In capable hands, the D3A divebomber could outmanoeuvre enemy fighters

Designed in 1936 for the Imperial Japanese Navy, the D3A, code named "Val" by the Allies, became one of the most effective dive bombers of World War II, sinking more Allied ships than any other enemy type in the first ten months of the Pacific War. In spite of the obsolete fixed landing gear, like that on the German Ju 87 Stuka, the Val was stable and easy to handle in a very steep dive, enabling the pilot to concentrate on his target.

With a bombing accuracy of over eighty per cent, the D3A did an enormous amount of damage at Pearl Harbor, 129 of them taking part in the attack. D3As went on to sink the British aircraft carrier HMS

Hermes and the cruisers *Cornwall* and *Dorsetshire* in April 1942. It was a crucial cog in Japanese strategy during the Battles of the Coral Sea and Midway.

In spite of excellent manoeuvrability, which enabled it to dogfight with enemy fighters, the Val was hopelessly vulnerable by 1943 and were shot down in great numbers. Aircrew quality had declined drastically as well, the veteran crews of 1941 replaced by hastily-trained novices. Bombing accuracy fell to under ten per cent and production was terminated in 1944.

The total production run included 1,495 D3As1s and just over 1,000 improved D3A2s, which had a 1300 hp engine. Some dedicated trainer aircraft were produced after the 'Val' left frontline service. Most of those left were overloaded with bombs and used as kamikazes through to the end of the war in 1945.

Specification (D3A1)

Powerplant: Mitsubishi Kinsei 44 1,075 hp 14-cylinder radial.
Dimensions: length 10.2 m (33 ft 5.5 in); height 3.35 m (11 ft); wing span 14.365 m (47 ft 1.5 in)
Weights: empty 2,408 kg (5,309 lb); operational 3,650 kg (8,047 lb).
Performance: maximum speed 389 km/h (242 mph); service ceiling 9,500 m (31,170 ft); range 1,820 km (1,131 miles).
Armament: two fixed 7.7 mm and one rear flexible machine guns; under fuselage 250 kg (550 lb) bomb; two 30 kg (66 lb) underwing bombs.

Kawanishi N1K1 Shiden (Violet Lightning)

The Shiden, Allied code-name 'George' was handicapped by the unreliability of its engine

Developed from the powerful N1K1 Kyofu floatplane fighter, the N1K1-J Shiden, Allied code name George, turned out to be one of the finest Japanese fighters of the war, flying for the first time in July 1943. With a massive radial engine and automatic manoeuvring flaps, the Shiden was the equal of any Allied fighter in the Pacific Theatre. Unfortunately for its Japanese Navy pilots, the engine was unreliable, the mid-wing configuration restricted visibility and the landing gear was so weak it tended to collapse with even moderate side loads. This lead to the low wing N1K2-J Shiden-Kai with a new tail and a simplified structure but with the old engine. It entered service in May 1944.

Even with a questionable powerplant, in the hands of a moderately trained pilot the George was something to be reckoned with. Plans were made for massive production by four companies and another four Navy arsenals but nothing ever came of the programme due to effective Allied bombing and submarine interdiction. In the end just over 1,000 Shidens and 400 Shiden-Kais were built. As with all other Japanese aircraft, they were used as kamikazes, far removed from their original mission.

Specification (N1K1-J)

Powerplant: Nakajima Homare 21 1,990 hp 180-cylinder radial.
Dimensions: length (1-J) 8.88 m (29 ft 1.75 in), (2-J) 9.35 m (30 ft 8.25 in); height (1-J) 4.058 m (13 ft 3.75 in), (2-J) 3.96 m (13 ft), wing span 12.0 m (39 ft 4.5 in)
Weights: empty (1-J) 2,897 kg (6,387 lb), (2-J) 2,657 kg (6,299 lb); operational (1-J) 4,321 kg (9,526 lb), (2-J) 4,860 kg (10,714 lb).
Performance: maximum speed (1-J) 583 km/h (362 mph), (2-J) 594 km/h (369 mph); service ceiling (1-J) 12,100 m (39,700 ft), (2-J) 10,750 m (35,400 ft); range (1-J) 1,430 km (989 miles), (2-J) 1,720 km (1,069 miles).
Armament: (1-J) four 20 mm cannon (two in wings, two in underwing pods), two 7.7 mm machine guns; (2-J) four 20 mm cannon in wings, two 250 kg (550 lb) underwing bombs or six rockets under fuselage.

Kawasaki Ki-45 Toryu (Dragon Slayer)

The Ki-45 (Allied name "Nick") was developed into a deadly night-fighter for the defence of Japan

Initially underpowered, when the Ki-45 first flew in January 1939 it was quite a change for the Imperial Japanese Army Air Force as the service's first twin engine fighter. As with most Japanese service aircraft, it was very manoeuvrable but, unlike so many others, the larger fuselage allowed continual modification for a variety of roles so it never fulfilled its initial function as a long range escort. A variety of armament packages, using anything from 7.7 mm machine guns to 37 mm (standard) and 75 mm cannon, were installed in several different configurations making the Toryu everything from an interceptor to an anti-shipping and ground attack aircraft.

As a night fighter and interceptor the Toryu began to live up to its name by continually downing B-29s from June 1944 on. Many used a fixed upward firing twin machine gun or cannon installation and the results were so effective the Germans adapted it for their night fighters, a very unusual reversal of the normal trend between the two Axis allies. The Ki-45 has the distinction of having made the first intentional suicide (later named kamikaze) attack on 27 May 1944, during the New Guinea campaign. The Toryu remained an effective and much used aircraft through to the end of the war, one of the few twin-engine fighters to hold its own in combat.

Specification (Ki-45-Kai-C)

Powerplant: twin Mitsubishi Ha-102 (Type-1) 1,080 hp 14-cylinder radials.

Dimensions: length 11 m (36 ft 1 in); height 3.7 m (12 ft 1.5 in); wing span 15.02 m (49 ft 3.5 in).

Weights: empty 4,000 kg (8,820 lb); operational 5,500 kg (12,125 lb).

Performance: maximum speed 550 km/h (340 mph); service ceiling 10,000 m (32,800 ft); range around 1,620 km (1,000 miles).

Armament: (night fighter) two 12.7 mm machine guns at 30 deg upward firing position, two 12.7 mm machine guns and one 20 mm or 30 mm cannon in the nose; (ground attack) two 250 kg (550 lb) bombs.

Kawasaki Ki-61 Hien (Swallow) "Tony"

Initially mistaken for a Messerschmitt Bf 109 by US pilots, the Ki-61 was a highly capable fighter

Based around a licence-built version of the Daimler-Benz DB 601 engine, the Ki-61-1 flew for the first time in December 1941 with initial deliveries in August 1942 and the first combats over New Guinea in April 1943. Since the Americans were expecting a squadron of Japanese-flown Bf 109s, the first encounter reports claimed Ki-61s were Messerschmitts, then an Italian design, which led to the Allied code name Tony.

The Hien was unique among Japanese aircraft: the only liquid-cooled, inline-engined fighter flown by either the Imperial army or navy. As a result, once Allied intelligence had properly identified it, the aircraft was easy to spot in combat.

The Ki-61-II, introduced in 1944, featured longer wings, a redesigned canopy and a very unreliable engine, typical of late war Japanese manufacturing. The -IIa brought back the original wing but most of the problems were solved with installation of a radial engine, changing the designation to Ki-100. This hasty transformation broke all the rules and was considered the finest of all Japanese fighters by many, combining excellent power with sterling manoeuvrability as well as the high altitude performance to intercept B-29s. Fortunately for USAAF crews, bombing attacks on the Japanese factories led to a terminal shortage of engines and other parts, so many Ki-61s remained incomplete at the end of the war.

Specification (Ki-61-IIb)

Powerplant: Kawasaki Ha-140 1,420 hp 12-cylinder inline.
Dimensions: length 9.16 m (30 ft .5 in); height 3.7 m (12 ft 2 in); wing span 12 m (30 ft 4.5 in).
Weights: empty 2,855 kg (6,294 lb); operational 3,825 kg (8,433 lb).
Performance: maximum speed 610 km/h (379 mph); service ceiling 11,000 m (36,089 ft); range 1,600 km (1,100 miles).
Armament: two 12.7 mm machine guns, two 20 mm cannon, two 250 kg (550 lb) bombs.

Mitsubishi A6M Zero-Sen "Zeke"

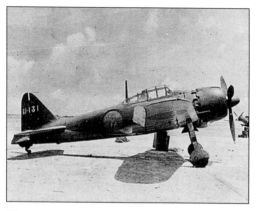

The astonishing performance of the Zero came as a nasty surprise to the Allies in 1941

Japan's most famous and most used aircraft, the Zero Fighter (Reisen or Zero-Sen) entered combat over China in July 1940 and stayed in first line service until the very end of World War II. At Pearl Harbor and in the early days of the war Jiro Horikoshi's design so outperformed every Allied fighter that all Japanese fighters were identified as Zeros. They seemed to be everywhere shooting down everything in spite of unheeded pre-war warnings from Americans

Claire Chennault in China and Tokyo US Navy attaché Stephen Jurika. Unfortunately, the Zero (code named Zeke) got its phenomenal manoeuvrability and range from light construction, no armour for the pilot and no self-sealing fuel tanks.

By 1943 the Zero was on the defensive, overpowered by superior fighters and better trained pilots who had no trouble in setting it on fire or chopping it apart. Yet the very mention of its name continued to get a fighter pilot's attention. Unable to remove it from production due to low industry output, the Japanese continued to refine the Zero and assign it to line squadrons, knowing most would be hacked from the sky. From the total production run of 10,937 many were used as kamikazes and over 300 were built as floatplane fighters, code named Rufe.

Specification (A6M2)

Powerplant: Nakajima NK1C Sakae12 925 hp 14-cylinder radial.

Dimensions: length 9.06 m (29 ft 9 in); height 2.92 m (9 ft 7 in); wing span 12 m (39 ft 4.5 in).

Weights: empty 1,680 kg (3,704 lb); operational 2,410 kg (5,313 lb).

Performance: maximum speed 509 km/h (316 mph); service ceiling 10,300 m (33,790 ft); range 3,110 km (1,940 miles).

Armament: two 20 mm cannon, two 7.7 mm machine guns, two 60 kg (132 lb) bombs.

Mitsubishi J2M Raiden (Thunderbolt)

A J2M Raiden, Allied codename "Jack", captured in Manila, the Philippines, March 1945

In a complete departure from the Japanese design philosophy of manoeuvrability over all else, the Raiden, known to the Allies as Jack, was optimized for performance, basically speed and rapid climb. A laminar flow wing was combined with excellent streamlining but when the first prototype flew in March 1942 it was plagued with faults, so the aircraft was redesigned but that failed to solve all the problems. In spite of a series of crashes the aircraft went into service as the J2M2 in December 1943, fitted with the same armament as the A6M2 Zero-Sen. The next version, the J2M3, was refitted with four 20 mm cannon, giving the aircraft the punch it needed. First combat operations were undertaken in September 1944.

Engine problems were finally sorted out with the J2M5 which was quite an effective B-29 killer in the last year of the war. Able to climb at over 3,000 feet per minute up to almost 40,000 feet, the Raiden could intercept Superfortress raids without much effort. Even though performance over handling was the primary requirement, pilots found the J2M had an excellent feel on the controls with no vicious habits, even in accelerated stalls. Just over 500 Raidens were produced by the time the war ended.

Specification (J2M3)

Powerplant: Mitsubishi MK4R-A Kasei 23a 1,820 hp 14-cylinder radial.
Dimensions: length 9.7 m (31 ft 9.75 in); height 3.81 m (12 ft 6 in); wing span 10.8 m (35 ft 5.25 in).
Weights: empty 2,574 kg (5,675 lb); operational 3,435 kg (7,573 lb).
Performance: maximum speed 612 km/h (380 mph); service ceiling 12,100 m (38,380 ft); range 1,055 km (655 miles).
Armament: four 20 mm cannon, two 60 kg (132 lb) bombs.

Mitsubishi G3M "Nell"

G3Ms flying from Indochina sank the British battleships Prince of Wales and Repulse in 1941

Unknown to western analysts, who considered the Japanese only imitators, when the prototype Ka-15 flew in July 1935 it was ahead of anything in the world except the Boeing Model 299 (later the B-17). With initial deliveries to the Navy in late 1936, the G3M was tested in combat over China, demonstrating a phenomenal range of over 1,200 miles by hitting targets from bases in Taipei. In spite of reports from some astute western military observers, when the Pacific War broke out Japanese land based bombers were assumed to have been launched from carriers. This was the only way they could explain how the British Prince of Wales and Repulse were sunk on 10 December 1941 when all Japanese airpower was supposedly out of range.

In spite of its success the Nell, as it was known to the Allies, was outclassed by the beginning of the war and was replaced by the more capable G4M. By 1943 most were phased out and turned into bomber crew transition trainers. A final version of the airframe was produced as the L3Y transport, code name Tina. The smooth, stressed-skin bomber remained a favourite of the crews that flew it.

Specification: (G3M3)

Powerplants: twin Mitsubishi Kinsei 51 1,300 hp 14-cylinder radials.

Dimensions: length 16.45 m (53 ft 11.5 in); height 3.685 m (12 ft 1 in); wing span 25 m (82 ft .25 in).

Weights: empty 5,243 kg (11,551 lb); operational 8,000 kg (17,637 lb).

Performance: maximum speed 415 km/h (258 mph); service ceiling 10,280 m (33,730 ft); range 6,228 km (3,871 miles).

Armament: one 20 mm cannon in dorsal fairing, three 7.7 mm machine guns in fuselage positions and dorsal turret, 800 kg (1,764 lb) bomb load or torpedo.

Mitsubishi G4M "Betty"

A captured G4M, Allied code-name 'Betty' under evaluation by the US Air Force

The major Japanese bomber of World War II, the G4M entered service over China with the Imperial Japanese Navy in April 1941. Under the same mandate as its predecessor, the G3M, to achieve maximum range (in this case a minimum 2,000 nautical miles with a full bomb load), the G4M's designers saved weight in every way possible. The goal was achieved but, as with Mitsubishi's famous Zero fighter, construction was light and there was very little protection for crew or fuel. Originally, four engines were called for but the Navy would not approve of the configuration.

The Betty (Allied code name) did indeed reach across the Pacific and, when not bothered by fighters, was very effective ranging far from its bases at will. Unfortunately for her crews, the bomber was dubbed "the flying cigarette lighter" by Allied pilots since it could be set ablaze with a minimum of ammunition. This led eventually to the G4M3 which reduced fuel from the original 5,000 to 4,400 litres in order to install armour plate and self-sealing fuel tanks but it was too late. The bomber was already past its prime, as was the case with most Japanese designs in quantity production during the war. In order to keep their air forces supplied with enough aircraft, Japanese planners never made enough room on the production lines for superior types. The Betty and the Zero were the central examples of this policy.

Specification (G4M2)

Powerplants: twin Kasei 22 1,850 hp 14-cylinder radials.
Dimensions: length 19.63 m (64 ft 4.75 in); height 4.11 m (13 ft 5.75 in); wing span 24.89 m (81 ft 7.75 in).
Weights: empty 8391 kg (18,500 lb); operational 12,500 kg (27,550 lb); maximum overload 15,000 kg (33,070 lb).
Performance: maximum speed 455 km/h (283 mph); service ceiling 9,100 m (30,000 ft); range (G3M1) 5,040 km (3,132 miles).
Armament: three flexible 7.7 mm machine guns in nose, dorsal, ventral positions, one 20 mm cannon in tail, one 7.7 mm gun in electric top turret; 1,000 kg (2,205 lb) bomb load or one 800 kg (1,764 lb) torpedo external.

Mitsubishi Ki-21 "Sally"

The Ki-21 was the main Japanese army bomber in 1941, but it was very vulnerable to fighters

The Ki-21 entered Japanese Army Air Force service in 1937 and was sent into combat over China almost immediately. The excellent results endeared the aircraft to crews and planners alike, though the bomber was revamped to conform to the lessons of combat. Among other things, defensive armament was changed from three to five or six 7.7 mm machine guns, the internal bomb capacity was increased, more armour was installed, fuel capacity was enlarged and the crew was expanded to seven.

When the Pacific War began, the aircraft (code named Sally) was the Army's primary bomber. Though it went on to fight on all fronts it was woefully ill-equipped to wage a modern war since even obsolete Allied fighters found the Sally a relatively easy target. Even so, Japanese crews were required to use the Ki-21 regardless of the opposition, a sad commentary on the short-sightedness of higher echelon staff in home islands. By the end of 1943 the Sally was being pulled from the combat zones to serve as a bomber transition trainer. The final version, as with the company's navy G3M Nell, was turned into the Ki-57 Topsy transport in order to get some use out of the existing production line.

Specification (Ki-21-II)

Powerplants: twin Mitsubishi Ha-101 1,490 hp 14-cylinder radials.
Dimensions: length 16 m (52 ft 6 in); height 4.85 m (15 ft 11 in); wing span 22.5 m (73 ft 9.75 in).
Weights: empty 6,070 kg (13,382 lb); operational 9,710 kg (21,395 lb).
Performance: maximum speed 478 km/h (297 mph); service ceiling 10,000 m (32,800 ft); range 2,200 km (1,370 miles).
Armament: six 7.7 mm machine guns; 1,000 kg (2,205 lb) bomb load.

Mitsubishi Ki-46 "Dinah"

Seen here under test by the USAF, the Ki-46 was designed for high altitude reconnaissance missions

From its first flight in November 1939 the Ki-46 proved one of the most successful of Japan's wartime aircraft, a favourite with pilots. Like other high speed, high altitude twin-engine aircraft on all sides (notably the British De Havilland Mosquito), it could perform its mission with some assurance of evading enemy fire. Initially used as a reconnaissance platform, the Ki-46 (code name Dinah) with its sleek airframe, again like the Mosquito, had excellent growth potential so it was turned into an interceptor, fighter-bomber and night fighter.

Subsequent versions proved able to do just about any job assigned in army and navy squadrons and crews loved their aircraft, regardless of the task. Later models had a streamlined canopy faired into the nose, replacing the earlier stepped arrangement, yielding a bit more speed and a more pleasing shape. Even late-war US fighters, operating with the benefit of radar control were not able to intercept the fast, high-flying Ki-46-III, which was also modified to become a fighter. One of the few aircraft able to fly high and fast enough to engage B-29s, the Ki-46-III-Kai carried two forward firing 20 mm cannon and one 37 mm cannon which fired upward at a 30 degree angle. This combination of performance and firepower was deadly. However, the design appeared too late, and pilot quality had declined too far for it to have a major impact on the Pacific air war.

The Dinah was one of the few pre-war Japanese designs which improved through the war and was able to match the Allied opposition. The Japanese army received over 1,700 Ki-46s of all versions.

Specification (Ki-46-III)
Powerplants: twin Mitsubishi Ha-112-II 1,500 hp 14-cylinder radials.
Dimensions: length 11 m (36 ft 1 in); height 3.88 m (12 ft 8.75 in); wing span 14.7 m (48 ft 2.75 in).
Weights: empty 3,831 kg (8,446 lb); operational 5,724 kg (12,620 lb).
Performance: maximum speed 630 km/h (391 mph); service ceiling 11,000 m (36,000 ft); range 4,000 km (2,485 miles).
Armament: none.

Nakajima B5N "Kate"

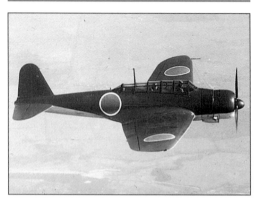

The B5N was the Japanese navy's main torpedo-bomber in the Pacific carrier battles of 1942

Ahead of its day for a 1935 design, the B5N carrier-based bomber (level and torpedo) was on the verge of obsolescence by 1941, yet it was a terrible surprise to the Americans at Pearl Harbor, even after being in combat over China since 1937. With a shallow depth Long Lance torpedo, supposedly impossible to develop, retractable landing gear, folding wings, a controllable pitch propeller, a streamlined cowling and a number of other innovations, it was devastating to the US Navy fleet. During the great Pacific naval battles of 1942, Kates, as they were known to the Allies, were pivotal in sinking four American carriers. The US Navy equivalent, the TBD Devastator, was hopelessly outclassed by the B5N in every area.

By the end of 1942 the relatively slow Kate, lacking armour plate and self-sealing fuel tanks, was easy meat for prowling Allied fighters and keen anti-aircraft gunners. Certainly part of the reason was the vulnerability of making long, level torpedo runs at heavily defended ships, something the Americans had discovered as well. As with most pre-war Japanese aircraft, the B5N was forced to stay in combat well into 1944 due to a lack of more modern types to replace it. During the final year of combat it became a kamikaze.

Specification (B5N2)

Powerplant: Nakajima Sakae 21 1,115 hp 14-cylinder radial.
Dimensions: length 10.3 m (33 ft 9.5 in); height 3.7 m (12 ft 3.75 in); wing span 15.52 m (50 ft 11 in).
Weights: empty 2,270 kg (5,024 lb); operational 3,800 kg (8,378 lb).
Performance: maximum speed 378 km/h (235 mph); service ceiling 7,640 m (25,000 ft); range 980-1,990 km (609-1,237 miles).
Armament: two 7.7 mm flexible rear machine guns, two 7.7 mm forward firing guns, centerline 800 kg (1,764 lb) torpedo or three 250 kg (550 lb) bombs.

Nakajima Ki-43 Hayabusa (Peregrine Falcon)

The Ki-43 (Allied codename "Oscar") was exceptionally manoeuvrable, but lightly-built and poorly armed

After entering production in March 1941 the Army Ki-43 was quickly confused with the Navy Zero-Sen by most outside Japan. Lighter, more manoeuvrable and far simpler to build and maintain, the Hayabusa became the JAAF's most numerous fighter, fighting on all fronts, and was code named Oscar. Even though it could outmanoeuvre any opponent the early Ki-43, like the Zero, did not have armour or self-sealing fuel tanks. Even worse, throughout the production run it had poor armament until the very last version which was pushed out in late 1944, long after the line should have been shut down.

In spite of its drawbacks, in the hands of a good pilot the Hayabusa could sting and almost all the JAAF's aces achieved most of their kills in it. The Ki-43-IIb had its wings clipped, resulting in an even better roll rate. Allied pilots often reported that "Oscar-type Zeros" were difficult targets to hit but once hit even slightly with .50 calibre fire they came apart or burned. By the end of the war almost 6,000 were built and many were thrown into the maw of last ditch kamikaze missions against the overwhelming American fleet.

Specification (Ki-43-II)

Powerplant: Nakajima Ha-115 1,105 hp 14-cylinder radial.
Dimensions: length 8.92 m (29 ft 3.75 in); height 3.27 m (10 ft 8.75 in); wing span (-IIb) 10.83 m (35 ft 6.75 in).
Weights: empty 1,750 kg (3,821 lb); operational 2,655 kg (5,850 lb).
Performance: maximum speed 515 km/h (320 mph); service ceiling 11,215 m (36,800 ft); range (with two drop tanks) 3,000 km (1,864 miles).
Armament: two 12.7 mm machine guns, two 250 kg (550 lb) bombs.

Nakajima Ki-44 Shoki (Demon) "Tojo"

This Ki-44 captured in the Philippines was later restored to flying condition by the US Air Force

In the same way the Navy realized it needed a higher performing aircraft in the J2M, the Army put out specifications for a Ki-43 replacement with speed and climb rate more important than manoeuvrability. The result was the Ki-44, which first entered production in May 1942. Even though the Shoki was superior to everything tested against it, pilots hated being blind over the nose, the faster landing speeds and the restrictions on severe manoeuvring. Yet, the Ki-44 could match Allied types in climbs and dives, giving pilots far more flexibility when engaging in combat, and the armament was far superior to the older Hayabusa. Shoki pilots could take on bombers or fighters without great worry.

These important characteristics made the fighter a good B-29 killer, one of the Japanese high command's major priorities during the last year of the war. With an initial climb rate of around 4,000 feet (1,200m) per minute the Tojo, as it was coded by the Allies, was an outstanding interceptor even though its top speed was not much more than 375 mph (603 km/h). Allied pilots who encountered Tojos were respectful of its performance but poor pilot training often made them easy targets.

Specification (Ki-44-IIc)

Powerplant: Nakajima Ha-109 1,520 hp 14-cylinder radial.
Dimensions: length 8.75 m (28 ft 8.5 in); height 3.25 m (10 ft 8 in); wing span 9.5 m (31 ft).
Weights: empty 2,106 kg (4,643 lb); operational 2,770 kg (6,107 lb).
Performance: maximum speed 605 km/h (376 mph); service ceiling 11,200 m (36,745 ft); range 900 km (560 miles).
Armament: two 12.7 mm machine guns and two 40 mm cannon; two 100 kg (220 lb) bombs.

Nakajima Ki-84 Hayate (Gale) "Frank"

Seen here in USAF markings, the heavily-armed Ki-84 was the best Japanese fighter of World War II

When the Ki-84 entered combat over China in the summer of 1944 Allied fighter pilots immediately recognized the new enemy as an equal if flown by a moderately trained pilot. Code named Frank, the Hayate was generally regarded as the finest of all wartime Japanese fighters, able to handle any Allied fighter or intercept the high flying B-29. In spite of a troublesome engine and landing gear prone to buckle, the Ki-84 was loved by its pilots, whether as a fighter or a fighter-bomber. When Allied pilots had a chance to test one they found out why: it easily bested the P-

51H and the P-47N.

In spite of the problems, most due to poor manufacturing, when Hayate units moved to the Philippines they gave American Navy and Army pilots the most serious opposition of the campaign. The fighter was one of the very few from Japanese industry which retained sterling manoeuvrability while gaining the performance needed to meet the Allies on an equal basis. From the beginning armament was what it should have been for a modern fighter; four 12.7 mm machine guns or two 12.7 mm guns plus two 20 mm cannon or four 20 mm cannon or two 20 mm and two 30 mm cannon. By any standards such a potent variety of weapons made the aircraft lethal even during brief encounters.

Specification (Ki-84-lc)

Powerplant: Nakajima Homare Ha-45 Model 11 1,900 hp 18-cylinder radial.

Dimensions: length 9.92 m (32 ft 6.5 in); height 3.38 m (11 ft 1.25 in); wing span 11.24 m (36 ft 10.5 in).

Weights: empty 2,680 kg (5,864 lb); operational 3,750 kg (8,267 lb).

Performance: maximum speed 624 km/h (388 mph); service ceiling 10,500 m (34,450 ft); range (with drop tanks) 2,920 km (1,815 miles).

Armament: two 20 mm and two 30 mm cannon, two 250 kg (550 lb) bombs.

Yokosuka D4Y Suisei (Comet) "Judy"

A Yokosuka D4Y3 under evaluation by No 16 Technical Air Intelligence Unit, USAF

First flown in November 1940, the D4Y dive bomber was designed around a licence-built DB 601 inline engine to match the range and speed of the Zero-Sen. Entering combat as a reconnaissance platform during the Battle of Midway, the Judy (Allied code name) had engine problems from the start, so subsequent versions were given radial engines. In the hands of a good crew (something that became increasingly difficult to produce in wartime Japan), the Suisei was indeed a potent aircraft, serving in the night fighter role (with two upward firing 20 mm cannon) as well. Indeed, it was able to match the Zero's top speed but when most of the Japanese carrier force was sunk by 1943 units operated from land bases without extensive fighter escort.

In spite of a relatively small internal bomb load (one 250 kg bomb), accuracy was the primary benefit of dive bombing and the D4Y had excellent stability in a steep dive, enabling pilots to hit their targets. Even so, increasing numbers of Allied fighters rendered the Judy ineffective on most occasions. When the D4Y4 single seat kamikaze was built with an 800 kg bomb load it became a fearsome weapon. Just over 2,000 D4Ys were built.

Specification (D4Y3)

Powerplant: Mitsubishi Kinsei 62 1,560 hp 14-cylinder radial.
Dimensions: length 10.22 m (33 ft 6.5 in); height 3.74 m (12 ft 3.25 in); wing span 11.5 m (37 ft 9 in).
Weights: empty 2,501 kg (5,512 lb); operational 4,657 kg (10,267 lb).
Performance: maximum speed 574 km/h (356 mph); service ceiling 10,500 m (34,500 ft); range 1,530 km (945 miles).
Armament: two 7.7 mm forward firing machine guns, one 7.7 mm flexible gun, one 250 kg (550 lb) internal bomb, two 30 kg (66 lb) underwing bombs.

Arado Ar 234 Blitz (Lightning)

The single-seat Ar 234 jet bomber proved almost impossible for Allied fighters to intercept

The world's first jet bomber, the Ar 234 was an outstanding aircraft in all respects, conceived in 1940 as a reconnaissance platform. Initially fitted with a jettisonable take-off dolly and skids for landing, the Blitz was quickly converted to retractable tricycle landing gear and built as the Ar 234B. The aircraft's first combat mission, a reconnaissance sortie, was flown in France on 20 July 1944 to determine Allied strength after the Normandy invasion. For the remainder of the war Blitz pilots flew high over Britain and the Continent, photographing Allied positions with impunity, yielding excellent intelligence which was useless for a near defeated army.

The first bomber sorties were flown in February 1945 but bombing at high speed was a much more complicated proposition than taking pictures at high altitude. Blitz pilots, along with their Me 262A-2 bomber pilots, found that releasing bombs in a shallow, high speed dive took a great deal of practice. Even with some success, the Ar 234 was finally made useless due to engine turbine blade failure, lack of fuel and prowling Allied fighters. Several four engine prototypes were built, some pressurized, and an experimental night fighter unit began operations in March 1945. The Ar 234, along with the Me 262, was a window on the future.

Specification (Ar 234B-2)

Powerplants: twin Junkers Jumo 004G axial-flow 900 kg (1,980 lb) static thrust axial-flow turbojets.

Dimensions: length 12.64 m (41 ft 5.5 in); height 4.29 m (14 ft 1.5 in); wing span 14.4 m (46 ft 3.5 in).

Weights: empty 5,200 kg (11,460 lb); operational 9,465 kg (11,460 lb).

Performance: maximum speed 742 km/h (461 mph); service ceiling 10,000 m (32,810 ft); range 1,560 km (970 miles).

Armament: no defensive armament, one 500 kg (1,100 lb) centerline bomb.

Dornier Do 17/215/217

The Dornier 217E-4 introduced BMW 801C engines and cable cutters on the wing leading edges

Dornier's Do 17 "Flying Pencil" earned its nickname honestly. A rejected 1934 civil airliner, the aircraft was turned into quite an effective medium bomber in the mid 1930s as the Luftwaffe was gaining strength. The first marked change in the aircraft came with the Do 17Z which had an enlarged nose to house the crew.

This later turned into the Do 215 export version. The bomber was a mainstay during the Battle of Britain and the Blitz although, as on most German daylight attacks, losses were heavy. Several Do 215Bs operated as night fighters/intruders with a solid gun nose.

Though the Do 217 looked very much like its older brothers, it was an entirely redesigned, larger aircraft with bigger engines and an increased bomb load. The Do 217E entered combat in late 1940 and was soon converted into a number of versions, including an effective night fighter and an air-to-ground guided missile launcher. A Do 217K sank the Italian battleship *Roma* with a 'Fritz X' wire-guided missile when the Italian naval forces in La Spezia sailed to join the Allies in September 1943

While the 217 was initially configured for the 215's DB 601 inline engines, when the BMW 801 radial was fitted it transformed the aircraft. The type remained a major Luftwaffe bomber and night fighter until the surrender.

Specification (Do 217E-2)

Powerplants: twin BMW 801A 1,580 hp 18-cylinder radials.

Dimensions: length 18.5 m (60 ft 10.5 in); height 5 m (16 ft 5 in); wing span 19 m (62 ft 4 in).

Weights: empty 8,850 kg (19,522 lb); operational 15,000 kg (33,070 lb).

Performance: maximum speed 515 km/h (320 mph); service ceiling 7,500 m (24,610 ft); range 2,100 km (1,300 miles).

Armament: one forward firing 15 mm machine gun in the nose, one 13 mm gun in the top turret, one 13 mm flexible gun in the ventral position, three 7.92 mm flexible guns in the nose and side windows; maximum bomb load 4,000 kg (8,818 lb).

Focke-Wulf Fw 190A/F/G

The outstanding Fw 190 enjoyed a marked superiority over Allied fighters from 1941-3

Chief company designer Kurt Tank's Würger (Butcher Bird) was Germany's most potent piston-powered World War II fighter. When the Fw 190A entered combat in the summer of 1941 it immediately outclassed the Spitfire V, which appeared sluggish and outdated by comparison. From that time on, in spite of some severe problems with the BMW 801 engine, the 190 kept even or ahead of Allied fighters through successive versions.

The Focke-Wulf was not only faster but its superior handling and faster roll rate gave it an edge in the hands of even less experienced pilots. Such sparkling performance combined with the 190's superior armament presented Allied pilots with a real challenge until German pilot training began to drop in quality.

The standard Fw 190A was quickly modified to perform a number of roles, particularly that of fighter-bomber in the F and G versions. These deleted the outer 20 mm cannon in favour of various combinations of bomb racks or cannon pods for the MK 103 30 mm cannon. Later versions of the FW 190A featured up to six 20 mm cannon (FW 190A-6R1); the A-6/R-6 had two 210 mm (8.27 in) unguided rockets with which to attack US heavy bombers.

The wide track landing gear assured ease of handling on take-off and landing, unlike the twitchy Messerschmitt 109. The 190 was also one of the first fighters to feature a clear rear canopy, allowing pilots to keep an excellent lookout for enemy fighters.

Specification (Fw 190A-8)

Powerplant: BMW 801D 1,700 hp 18-cylinder radial.
Dimensions: length 8.84 m (29 ft); height 3.96 m (13 ft); wing span 10.49 m (34 ft 5.5 in).
Weights: empty 3,200 kg (7,055 lb); operational 4,900 kg (10,800 lb).
Performance: maximum speed 653 km/h (408 mph); service ceiling 11,410 m (37,400 ft); range 900 km (560 miles).
Armament: two 13 mm machine guns plus four 20 mm cannon or two 20 mm and two 30 mm cannon.

Focke-Wulf Fw 190D/Ta 152

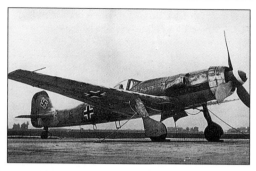

The extended wings of the Ta 152 gave it superb performance at high altitude

In honour of designer Kurt Tank, the Fw 190's designation was changed to Tank or Ta 152. This beautiful inline engined fighter was to be the ultimate version of the famous fighter but delays resulted in the stopgap Fw 190D, in itself an outstanding aircraft. In the chaotic final year of the Third Reich the D ended up being the major inline engine version with only a few Ta 152Hs, and possibly a few Ta 152Cs, getting into combat.

The new Junkers Jumo 213 powerplant made the aircraft, once again, the fastest Luftwaffe operational fighter and those pilots with the skill to use such advantages did very well. Unfortunately excellent fighter designs could not compensate for poor production standards, lack of fuel, poor pilot training and overwhelming Allied numerical superiority.

The extended wing (14.5m), high altitude Ta 152H was indeed a sterling performer with a top speed of 755 km/h (472 mph) and a service ceiling of 15,000 m (49,215 ft). It was armed with a 30 mm cannon in the nose and two 20 mm cannon in the wing roots. Had it been built in enough numbers and been flown by expert pilots it could have taken its place alongside the Me 262 as a near unbeatable air superiority fighter and bomber killer. The lower altitude version, the Ta 152C, barely made it out of the test phase before the war ended. Total Ta 152 production barely exceeded 200 aircraft.

Specification (Fw 190D-9)

Powerplant: Junkers Jumo 213A-1 1,776 hp 12-cylinder inline.
Dimensions: length 10.2 m (33 ft 5.25 in); height 3.35 m (11 ft .25 in); wing span 10.49 m (34 ft 5.5 in).
Weights: empty 3,500 kg (7,720 lb); operational 4,840 kg (10,670 lb).
Performance: maximum speed 704 km/h (440 mph); service ceiling 10,000 m (32,810 ft); range 900 km (560 miles).
Armament: two 13 mm machine guns, two or four 20 mm cannon, one 30 mm cannon.

Focke-Wulf Fw 200C Condor

Ranging far across the Atlantic, the Fw 200 inflicted heavy losses on Allied merchant shipping

Another pre-war airliner developed into a bomber, the Fw 200 had such impressive range that zealous planners made the mistake of demanding too much of it. From the start the landing gear and the fuselage were too weak to carry full military loads, leading to collapsed undercarriages and broken backs throughout its history. Nevertheless, from the time the Condor entered combat in June 1940, the patrol bomber became, according to Winston Churchill, the "scourge of the Atlantic" by sinking Allied shipping far out to sea where protective fighters could not reach it.

In spite of increasing Allied ability to intercept and down them, Fw 200s were continually modified for their role, finally carrying search radar and Hs 293 guided missiles in their hunt for shipping. Mounting losses finally forced the Luftwaffe to turn the aircraft back into its intended role as a transport aircraft. Even so, the small total production run of 276 (the last was built in February 1944) was far out of proportion to the aircraft's impressive lethality since it seemed to roam the Atlantic at will with great success.

The only version built, the Fw 200C, was flown with rare exception by one unit, KG 40, operating from airfields in France. The results far outweighed the size of the Condor effort, yielding one of Germany's most successful and least costly military aviation programmes.

Specification (Fw 200C-3)

Powerplants: four BMW-Bramo Fafnir 323R-2 1,200 hp 9-cylinder radials.
Dimensions: length 23.46 m (76 ft 11.5 in); height 6.3 m (20 ft 8 in); wing span 30.855 m (107 ft 9.5 in).
Weights: empty 12,951 kg (28,550 lb); operational 22,700 kg (50,045 lb).
Performance: maximum speed 360 km/h (224 mph); service ceiling 5,800 m (19,030 ft); range 3,550 km (2,206 miles).
Armament: one 15 mm, 20 mm or 7.92 mm gun in top turret, one 20 mm flexible cannon in front ventral gondola, three 7.92 mm flexible guns in rear ventral gondola plus two 7.92 mm or two 13 mm guns in side positions, one 13 mm flexible gun in rear dorsal position; maximum bomb load 2,100 kg (4,626 lb).

Heinkel He 111

The Heinkel He 111 medium bomber served the Luftwaffe throughout the war

Secretly test flown as a bomber in February 1935, the He 111 prototype was labelled an airliner to hide its purpose, the reverse of most German bomber development. By February 1937 the He 111B entered combat with the Kondor Legion in Spain and was quite effective, a success that would later cost the Luftwaffe dearly. As the major medium bomber in the Battle of Britain of summer 1940, the 111 was vulnerable to effective fighter opposition, as were all Luftwaffe bombers. More defensive guns and crew members were added until the aircraft's initial excellent performance was significantly downgraded.

Even though the He 111 was outmoded by early war standards, it was needed on all fronts. Production was standardized on the H model and the aircraft was used for a wide variety of missions, including guided missile and V-1 flying bomb launching. One variant, the He 111Z was actually two He 111Hs joined with a common wing centre section; it served to tow the ponderous Me 321 Gigant into the sky.

Unable to carry enough defensive guns in an already overweight aircraft, crews found themselves easy meat for Allied fighters. Under Goring's poor leadership, the Luftwaffe failed to develop a replacement, instead creating such monsters as the problem-plagued Heinkel He 177. The Spanish built a Merlin engined version through to 1956.

Specification (He 111H-6)

Powerplants: twin Junkers Jumo 211F-1 1,300 hp 12-cylinder inlines.

Dimensions: length 16.4 m (53 ft 9.5 in); height 4 m (13 ft 1.5 in); wing span 22.6 m (74 ft 2 in).

Weights: empty 8,680 kg (19,136 lb); operational 14,000 kg (30,864 lb).

Performance: maximum speed 440 km/h (273 mph); service ceiling 6,500 m (21,330 ft); range 2,300 km (1,429 miles).

Armament: one 20 mm cannon in front ventral gondola, six 7.92 mm machine guns in various positions; bomb load 2,000 kg (4,410 lb) or two torpedoes.

Heinkel He 162 Salamander

Like the first prototype, this He 162 crashed under evaluation by the RAF after the war, killing its pilot

A product of desperation, the He 162, also known as the Volksjäger (People's Fighter), flew only three months after the September 1944 specification for a cheap jet fighter was issued. The all-wood aircraft was to be produced in massive numbers and flown by relatively inexperienced pilots. As it turned out, flying any jet fighter was something best left to the experts, not to mention something like the 162 which had a number of severe problems requiring a deft hand on the controls. The fantastic quantities ordered: 1000 per month, rising to 4000 per month could never have been achieved either.

By the end of the war around 320 had been completed and another 800 were being built, a remarkable achievement considering the chaotic conditions of the Third Reich's last days.

One squadron (1 Gruppe/JG 1) had actually converted to the He 162 at Leck and began to see combat in mid-April 1945 with the first victory, and the first loss, occurring on the 19th. However, many more losses were due to accidents in trying to fly the jet. Heinkel test pilot Flugkapitan Peter was killed testing the He 162 in December 1944.

In tight turns parts tended to fly off and the aircraft would spin out of control, and the engine was prone to flame-outs. In the three weeks from April 13th to the end of the war 13 aircraft and 10 pilots were lost, only two or three in combat. Even so, when everything worked it was a delight to fly and was considered the best jet of the war as a gun platform with no directional snaking.

Specification (He 162A-2)

Powerplant: BMW 003 980 kg (2,028 lb) thrust axial-flow turbojet.
Dimensions: length 9 m (29 ft 8.25 in); height 2 m (6 ft 6 in); wing span 7.2 m (23 ft 7.50 in)
Weights: empty 1,760 kg (3,875 lb); operational 2,700 kg (6,184 lb).
Performance: maximum speed 910 km/h (562 mph); service ceiling 12,000 m (38,900 ft); range 620 km (385 miles).
Armament: two 20 mm cannon.

Heinkel He 219A Uhu (Owl)

The radar-equipped He 219 night fighter inflicted heavy losses on RAF heavy bombers over Germany

Originally proposed as a multirole fighter-bomber in August 1940, the He 219 drew no official interest until late 1941 when RAF heavy bomber attacks became serious enough to require an effective night fighter. The first flight was made in November 1942 and production began a year later. The aircraft was a major leap forward with speeds of over 400 mph (650 km/h), outstanding manoeuvrability and hard-hitting armament. When the first He 219 night fighter unit was formed at Venlo in Holland, several of the prototypes were considered capable enough to be sent for combat use. In the first six missions 20 bombers were reported as destroyed, including six Mosquitoes which had been almost invulnerable up to that time.

As was typical of mercurial wartime Luftwaffe leadership, the He 219 was ordered in multiple versions, each quite effective, but then production was halted. Those flying the 219, now a real thorn in the RAF's side, continued to send in sparkling reports on their fighter's lethality but official response never seemed to be consistent. By the time the war was over only 268 He 219As had been produced with plans for B and C models which would have reverted to the original multirole concept as fighter-bombers and destroyers. Had the 219 been given some priority it would have been a real threat.

Specification (He-219A)

Powerplants: twin Daimler-Benz DB 603G 1,900 hp 12-cylinder inlines.
Dimensions: length 15.54 m (50 ft 11.75 in); height 4.1 m (13 ft 5.5 in); wing span 18.5 m (60 ft 8 in).
Weights: empty 11,200 kg (24,692 lb); operational 15,200 kg (33,730 lb).
Performance: maximum speed 670 km/h (416 mph); service ceiling 12,700 m (41,660 ft); range 2,000 km (1,243 miles).
Armament: four 30 mm and two 20 mm forward firing cannon plus two 30 mm 65 deg upward firing Schräge Musik cannon.

Henschel Hs 123

The Hs 123 enjoyed a new lease of life as a light bomber on the Russian front

In spite of the apparent obsolescence of its 1935 open cockpit biplane design, the Hs 123 ground attack/dive bomber was not only deadly during the Spanish Civil War but held its own in World War II, particularly on the Russian Front. The real key to the aircraft's success was its delivery of immediate support to troops on the ground; it was slow and rugged enough to lay down accurate fire within yards of friendly positions. Even though production was halted before the invasion of Poland, the Hs 123 was so useful that Luftwaffe planners could not pull it out of combat. Flak seemed to have far less effect on it than other close air support aircraft, particularly the Ju 87, and pilots found its manoeuvrability ideal for evading not only ground gunners but enemy fighters.

Even though Hs 123 units were continually slated for conversion to newer types or disbandment, they kept flying in the worst conditions through the end of 1944. When the snow and freezing temperatures of the Balkans and Russia grounded other more modern aircraft, the Henschel kept on flying until, at last, every one of them was worn out and pushed aside. Its replacement, the heavily armed and armoured twin-engine Hs 129, was a worthy successor but there were never enough to pull its venerable older brother biplane out of combat.

Specification (Hs 123A-1)

Powerplant: BMW 132 Dc 880 hp 9-cylinder radial.
Dimensions: length 8.3 m (27 ft 4 in); height 3.2 m (10 ft 6.5 in); wing span 10.5 m (34 ft 5.5 in).
Weights: empty 1,504 kg (3,316 lb); operational 2,217 kg (4,888 lb).
Performance: maximum speed 345 km/h (214 mph); service ceiling 9,000 m (29,530 ft); range 850 km (530 miles).
Armament: two 7.92 mm forward firing machine guns; two 20 mm underwing cannon or four 50 kg (110 lb) bombs.

Junkers Ju 52/3m

The Ju 52 remained the Luftwaffe's main transport from 1939 until the final flights from Berlin in 1945

Though outmoded by World War II, the Junkers 52/3m, which flew for the first time in April 1932, became Germany's primary transport through every major campaign of the war. Nicknamed "Tante Ju" (Aunt Ju) by its crews, the aircraft was developed from the single-engine Ju 52 and served across the world as an airliner. In 1934 the Luftwaffe began to buy it as a heavy bomber but that role quickly gave way to transport of troops and supplies as other bombers took its place. When the Spanish Civil War began in July 1936 the Ju 52/3m was more than useful for moving troops, even though it still operated as a bomber at times.

When Germany invaded Poland in September 1939, the Junkers came into its own with some 550 in service. From that point on, Tante Ju was everywhere and she was particularly crucial in keeping German armies in Russia and North Africa supplied. Unfortunately, she was also a sitting duck at her 200 km/h (124 mph) cruising speed and great numbers were downed by Allied fighters, often carrying hundreds of troops to their death. Even so the Luftwaffe could not come up with anything to replace it, so its 1,845 kg (4,067 lb) payload was a critical part of Germany's wartime campaigns. After the war Spain continued to build the aircraft into the 1950s.

Specification (Ju 52/3mg4e)

Powerplants: three BMW 132A-3 525 hp 9-cylinder radials.
Dimensions: length 18.9 m (62 ft); height 4.5 m (14 ft 9 in); wing span 29.25 m (95 ft 11.5 in).
Weights: empty 6,510 kg (14,354 lb); operational 10,500 kg (23,157 lb).
Performance: maximum speed 270 km/h (168 mph); service ceiling 5,500 m (18,046 ft); range 915 km (568 miles).
Armament: one 7.9 mm machine dorsal gun, two 7.9 mm guns in side windows.

125

Junkers Ju 87 Stuka

The vertical-diving Ju 87 Stuka was deadly accurate but highly vulnerable to enemy fighters

This single weapon struck more terror into Germany's enemies than any other for the first three years of World War II. The Stuka (an abbreviation of Sturzkampfflugzeug, the word for dive bomber), which flew for the first time in 1935, was the result of Ernst Udet's obsession with dive bombing. In Spain, without serious enemy fighter opposition, the Ju 87 proved ideal for the mission, able to dive very steeply at relatively slow speeds with an automatic pull-out device, yielding excellent accuracy. The same proved true in the early campaigns of the war through to the fall of France. During the Battle of Britain the slow Stuka with its fixed landing gear was slaughtered by British fighters and was withdrawn to attack targets (convoys, tanks, troops) which were not heavily defended.

In spite of its drawbacks in a modern war, the Ju 87 got a new lease of life during the Russian campaign where it was an exceptional tank killer, carrying two 37 mm flak cannons in underwing pods. Hans-Ulrich Rudel, the most decorated Luftwaffe pilot of the war, claimed over 500 Russian tanks. Throughout the war Stuka production was repeatedly ordered to end, only to be restarted since there were very few aircraft so specialized to take its place. The Ju 87 remains one of the most famous of all World War II aircraft.

Specification (Ju 87B-1)

Powerplant: Junkers Jumo 211Da 1,100 hp 12-cylinder inline.
Dimensions: length 11.1 m (36 ft 5 in); height 3.9 m (12 ft 9 in); wing span 13.8 m (45 ft 3.25 in).
Weights: empty 2,750 kg (6,080 lb); operational 4,250 kg (9,371 lb).
Performance: maximum speed 390 km/h (242 mph); service ceiling 8,000 m (26,250 ft); range 600 km (373 miles).
Armament: two 7.92 mm forward firing machine guns, one 7.92 mm flexible gun in rear, one 500 kg (1,100 lb) bomb below fuselage, four 50 kg (110 lb) bombs under wings.

Junkers Ju 88

This Ju 88 was captured when its crew mistakenly landed in Devon, thinking they were over France

Yet another civil airliner converted into a bomber, the Ju 88 flew for the first time in December 1936 and was immediately regarded as the ideal Schnellbomber (high-speed bomber) around which to build the growing Luftwaffe. Entering service just days after the start of World War II, the Ju 88 quickly became an airframe modified for many different missions, used in more different roles than any single German wartime aircraft - bomber, fighter, night fighter, destroyer, tank buster, reconnaissance, dive bomber and just about anything else Berlin could think of.

The aircraft's original excellent performance gradually suffered as more was required of the airframe and as with all German bombers defensive armament was grossly inadequate. Nevertheless, even loaded the Ju 88 had enough performance left to fare far better than its fellow Dorniers and Heinkels, particularly in the fighter-bomber role. The Ju 88A remained the primary version through 1943 but new versions, particularly the Ju 88S, and then redesigned Ju 188s and 388s gave new life to the basic aircraft, pushing top speeds well over 400 mph (650 km/h). Almost 15,000 Ju 88s were built and the night fighter, which could carry an impressive combination of radar and weapons, was the most successful of all types in downing RAF bombers.

Specification (Ju 88A-4)

Powerplants: twin Junkers Jumo 211J 1,340 hp 12-cylinder inlines.
Dimensions: length 14.4 m (47 ft 2.25 in); height 4.85 m (15 ft 11 in), wing span 20.13 m (65 ft 10.5 in).
Weights: empty 8,000 kg (17,637 lb); operational 14,000 kg (30,865 lb).
Performance: maximum speed 433 km/h (269 mph); service ceiling 8,200 m (26,900 ft); range 1,790 km (1,112 miles).
Armament: two 7.92 mm forward firing machine guns, two to four 7.92 mm rearward firing flexible guns, plus two 7.92 mm flexible forward firing guns in later versions; 500 kg (1,100 lb) internal bomb load, six underwing racks for 1,000 kg (2,205 lb) bomb load.

129

Messerschmitt Bf 109

The Bf 109 was Germany's premier fighter aircraft in 1939 and served until the end of the war

When Willy Messerschmitt designed the 109 in 1934 for the Bayerische Flugzeugwerke (Bf), the small fighter captured the world's imagination from the start, a hold it has kept to the present day. With the Bf 109E the Luftwaffe established its superiority over most of Europe through to the summer of 1940 when it met the RAF Hurricane and Spitfire during the Battle of Britain. Hampered by being at the end of their range, German pilots did not have time to use many of the fighter's sterling qualities and they were unable to prevent RAF victory.

Even though the Fw 190 came along in 1941, the 109 kept up with Allied fighters through successive versions until the end of the war. A tricky aircraft on take-off and landing, the Messerschmitt required skill and experience. Most of Germany's leading aces, the experten, including all-time leader 352 victory ace Erich Hartmann, got their kills in 109s and preferred to fly them even when newer fighters were available.

When the American daylight bombing campaign was at its height, the 109 was lightened and used as a top cover fighter while heavier Fw 190s attacked the bombers, a testimony to the Messerschmitt's handling. Final versions included a high altitude fighter with a pressurised cockpit. After the war the 109 was built in Czechoslovakia, then in Spain until 1956. A total of some 35,000 examples were assembled, one of the largest production runs in history.

Specification (Bf 109G-6)

Powerplant: Daimler-Benz DB 605A-1 1,475 hp 12-cylinder inline.

Dimensions: length 9.04 m (28 ft 8 in), height 2.59 m (8 ft 6 in); wing span 9.92 m (32 ft 6.5 in).

Weights: empty 2,700 kg (5,900 lb); operational 3,400 kg (7,500 lb).

Performance: maximum speed 630 km/h (387 mph); service ceiling 11,600 m (38,500 ft); range 700 km (425 miles).

Armament: two 13 mm machine guns, one 20 mm cannon, two 20 mm cannon in underwing pods.

Messerschmitt Bf 110

After a poor showing in the Battle of Britain, the Bf 110 became a very effective night fighter

During the late 1930s several nations looked at the twin-engine fighter as a way of harnessing enough horsepower to propel a good gun platform with additional fuel for long range. With few exceptions, among them the P-38 Lightning and the Mosquito, the resulting aircraft were under-powered and vulnerable to single-engine opponents. The Bf 110, which flew for the first time in May 1936, was one of the failures as a pure fighter, getting shot to pieces during Battle of Britain. It had the range needed to protect the bombers but not the manoeuvrability to defeat single-engine fighters. Zerstörer (Destroyer), the name given to the aircraft's mission, was never a reality until 1942 when night fighters were desperately needed to stem the tide of RAF Bomber Command.

With the addition of multiple cannon and more powerful engines, the 110 became an extraordinary night fighter, serving until the end of the war. The Bf 110F-4a carried Lichtenstein radar which increased its ability to intercept RAF bombers at the cost of some reduction in performance.

With the failure of the Me 210, the Bf 110 was put back into production and developed as a heavy day fighter to engage US heavy bombers. The Bf 110G series was manufactured in a bewildering number of variants, the most heavily armed of which was the Bf 110G-2/R-1 with a 37 mm cannon.

Specification (Bf 110G-4/R3)

Powerplants: twin Daimler-Benz DB 605B 1,475 hp 12-cylinder inlines.

Dimensions: length 12.66 m (41 ft 6.75 in); height 4 m (13 ft 1.75 in); wing span 16.25 m (53 ft 4.75 in).

Weights: empty 5,200 kg (11,220 lb); operational 9,400 kg (20,700 lb).

Performance: maximum speed 555 km/h (342 mph); service ceiling 8,000 m (26,000 ft); range 2,100 km (1,305 miles).

Armament: two 30 mm, two 20 mm forward firing cannon, two 20 mm cannon firing obliquely upward, two 7.92 mm flexible rear firing machine guns.

Messerschmitt 163 Komet

The rocket-powered Me 163 Komet killed more of its own pilots than Allied aircraft

A vision of things to come, the rocket-powered Me 163 flying wing, the dream of designer Alexander Lippisch, first flew under power in August 1941. On October 2nd Heini Dittmar piloted the prototype well beyond the existing world speed record to 1004 km/h (624 mph) but the flight had to remain a secret. Redesigned as a point defence fighter with very limited range, the tiny Me 163B entered combat in May 1944 as one of many weapons which came too late to stop the American daylight bombing campaign. In spite of the explosive nature of the rocket engine's fuel, all 163s were wonderful aircraft to fly, light on the well harmonized controls, with a phenomenal climb rate of 5,000 m (16,000 ft) per minute. Nothing in the Allied inventory could catch it with its glide speed of 500 mph (805 km/h).

Unfortunately for its pilots, engine throttling was temperamental, often causing flame-outs if varied, so pilots were ordered to climb at full power until the fuel ran out after about 2.5 minutes. Then the pilot was to glide through the enemy bomber stream, getting what results he could, before gliding back to base where Allied fighters often waited for the helpless aircraft. Since the fighter dropped its wheeled trolley on take-off, it landed on a skid. Pilots had to be experts in high speed glider landings. The later Me 163C and Me 263 (with retractable landing gear) never got into service and Me 163Bs accounted for only 16 kills.

Specification (Me 163B-1)

Powerplant: Walter HWK 109-509A-2 1,700 kg (3,750 lb) thrust bi-fuel liquid rocket.

Dimensions: length 5.69 m (18 ft 8 in); height 2.74 m (9 ft); wing span 9.3 m (30 ft 7 in).

Weights: empty 1,908 kg (4,200 lb); operational 4,310 kg (9,500 lb).

Performance: maximum speed 955 km/h (592 mph); service ceiling 15,000 m (39,500 ft); range (radius of action) 40 km (25 miles).

Armament: two 30 mm cannon.

Messerschmitt Me 210/410 Hornisse

The heavily-armed Me 410 was built to destroy Allied heavy bombers attacking Germany

A variation on a theme, the Me 210 was designed in 1937 to succeed the Bf 110 as an escort fighter but the aircraft was a dismal failure and production was stopped in April 1942 after just over 350 had been built. Unwilling to let his reputation suffer so badly, Willy Messerschmitt redesigned the aircraft as first the 310 and then the 410. The improvements were deemed satisfactory enough to reopen the production line. After the new prototype flew in late 1942 the aircraft was quickly adapted to a number of roles, including reconnaissance, attack, bomber destroyer, tank killer and night fighter.

The enlarged fuselage, which included a bomb bay, was ideal for mounting from four to eight 20 mm cannon or even a 50 mm cannon, or variations which included 30 mm cannon and 210 mm rocket launchers. The heavy cannon and rockets reduced performance, but a single hit could bring down a B-17 or B-24. Pilots found the 410 handled far better than the 210, or the 110 for that matter, with excellent visibility and remotely sighted rear guns for self-protection. In spite of the improvements, as with most German twin-engine fighters, the aircraft was vulnerable to Allied fighters if speed was not maintained. At a top speed of 385 mph (620 km/h) the 410 could often get away from the opposition, particularly in a shallow dive.

Specification (Me 410A-1)

Powerplants: twin Daimler-Benz DB 603A 1,750 hp 12-cylinder inlines.

Dimensions: length 12.45 m (40 ft 11.5 in); height 4.3 m (14 ft .5 in); wing span 16.4 m (53 ft 7.75 in).

Weights: empty 6,150 kg (13,550 lb); operational 10,650 kg (23,500 lb).

Performance: maximum speed 620 km/h (385 mph); service ceiling 10,000 m (32,800 ft); range 2,330 km (1,450 miles).

Armament: four 20 mm forward firing cannon, two 7.92 mm forward firing machine guns, two 13 mm remotely-controlled rear firing guns.

Messerschmitt Me 262

Technical difficulties delayed the arrival of the outstanding Me 262 jet interceptor

Willy Messerschmitt's revolutionary Me 262 became the world's first operational jet aircraft with a top speed of 540 mph (870 km/h), making the newest piston engine types immediately obsolete. A preview of what was to come, lack of jet engines forced the company to test fly the airframe with a piston engine in April 1941. The first turbojet powered flight was made by the tailwheel-equipped Me 262V3 on 18 July 1942. Production (in modified form with a tricycle landing gear) was called for immediately but the first examples were not ready until May 1944. Hitler saw the aircraft as an ideal fast bomber so he ordered production centred around the Me 262A-2

Sturmvogel, which carried only two 250 kg (550 lb) bombs, rather than the Me 262A-1a Schwalbe. The first combat trials unit began to fly in June 1944 with the first bomber squadrons flying combat in September.

Though almost 2,000 262s were built, lack of hot section turbine blade technology (rather than Hitler's decision to build bomber versions) delayed operational use so that only a few squadrons got into combat. Even so, Me 262B night fighters and photo reconnaissance versions flew several missions before the war ended. When captured Me 262s were evaluated after the war the type's advanced design features, including swept wings, leading edge slats and flying tail, were incorporated into many new aircraft, particularly the F-86 Sabre.

Specification (Me 262A-1a)

Powerplants: twin Junkers Jumo 004B 900 kg (1,980 lb) thrust axial-flow turbojets.

Dimensions: length 10.6 m (34 ft 9.5 in); height 3.8 m (12 ft 7 in); wing span 12.51 m (41 ft 4.5 in).

Weights: empty 4,420 kg (9,740 lb); operational 6,396 kg (14,100 lb).

Performance: maximum speed 870 km/h (540 mph); service ceiling 11,500 m (37,565 ft); range 480 km (300 miles).

Armament: four 30 mm cannon, 24 R4M unguided underwing rockets.

PZL P.11

Outnumbered and out-gunned, Polish P.11s put up gallant resistance in 1939

When the gull-winged PZL (National Aircraft Factory) P.11 first flew in August 1931 it was quite a modern aircraft. Subsequent modifications kept it effective well into the late 1930s when it made up the bulk of Polish Air Force fighter units. Several were exported to Bulgaria, Greece and Rumania.

By the time of the German invasion in September 1939 most frontline P.11c fighters in twelve squadrons still had a woefully inadequate armament of two 7.7 mm machine guns, though four could be fitted. To add to the problems, very few were equipped with radios, a necessity of modern war, particularly if fighters are to be directed to incoming enemy formations.

In spite of these enormous drawbacks, facing a very modern, numerically superior enemy, Polish P.11 pilots managed to claim 126 German aircraft for the loss of 114 of their own. Certainly this was not a favourable kill ratio, particularly in the face of a larger enemy force, but that it was better than one-to-one says a great deal about both pilots and planes. When the German victory was no longer in doubt, many Polish pilots flew their aircraft to neutral countries and made their way to the west, to continue the fight. The skill and tenacity of Polish pilots would be well used in RAF Polish squadrons during the rest of the war. Most of them had learned their skills in the P.11.

Specification (P.11c)

Powerplant: PZL Mercury VIS2 645 hp 9-cylinder radial.
Dimensions: length 7.55 m (24 ft 9 in); height 2.85 m (9 ft 4 in); wing span 11.48 m (35 ft 2 in).
Weights: empty 1,146 kg (2,524 lb); operational 1,798 kg (3,960 lb).
Performance: maximum speed 389 km/h (242 mph); service ceiling 11,000 m (36,090 ft); range 809 km (503 miles).
Armament: four 7.7 mm machine guns; external bomb load two 12 kg (27 lb) bombs.

Ilyushin Il-2 Shturmovik

The heavily-armoured Il-2 Shturmovik spearheaded Russian ground attack missions

The "Flying Tank" of the Soviet Air Force, the Il-2 has the distinction of being the most produced aircraft in history at a total of 36,163. After the prototype was flown in December 1939, the production line was quickly opened and the aircraft entered combat in July 1941, just after the German invasion. From that point on the Shturmovik (armoured attacker) became a highly effective ground attack aircraft through to the end of the war with successive versions being fielded as late as the Korean War. Surrounded by armour 4 mm to 8 mm thick, the aircraft's pilot, engine and fuel tanks were difficult to hit from above and below, making the Il-2 near impervious to fighters and flak. Such rugged construction, combined with several cannon and low-level flying, made the aircraft a pivotal piece of flying artillery in opposing the German advance, then throwing it back.

The initial single-seat versions proved vulnerable, so a second seat was added for a rear-firing gunner. Later versions were fitted with two 37 mm cannon which could punch through most tank armour. The standard Soviet nozhnitsi (scissors) tactic decimated German columns and trains by flying pairs or line abreast Il-2 formations in opposing, cross over low-level S-turns, concentrating firepower while confusing anti-aircraft gunners and fighters.

Specification (Il-2m3)

Powerplant: Mikulin AM-38F 1,750 hp 12-cylinder inline.
Dimensions: length 11.65 m (38 ft 2.7 in); height 3.4 m (11 ft 1.9 in); wing span 14.6 m (47 ft 10.75 in).
Weights: empty 4,525 kg (9,976 lb); operational 6,160 kg (13,580 lb).
Performance: maximum speed 372 km/h (231 mph); service ceiling 6,500 m (21,330 ft); range 765 km (475 miles).
Armament: two 23 mm cannon, two 7.62 mm machine guns, one 12.7 mm rearward firing gun; eight 82 mm or four 132 mm rockets, two 250 kg (550 lb) or four 100 kg (220 lb) bombs.

Ilyushin Il-4

This Il-4 was captured intact by the Finns who used it against its former owners

Initially produced as the DB-3 (the Soviet designation for long range bomber) in 1937, this Russian long-range bomber proved able to fly tactical as well as strategic bombing missions. Though western bombers were better known for hitting Germany, the Il-4 did raid Berlin and other targets in Germany as well with a bomb load equal to the B-17. The first Soviet air attack on the German capital took place only six weeks after the invasion of Russia. It could not be repeated until 1944 when the great Soviet summer offensive drove the Germans back through western Russia and across Poland, recapturing the pre-war bases of the Soviet bomber regiments.

The Il-4's main drawback was a woefully inadequate three gun defensive armament, something that changed to only five guns by the end of the war. Even so, the Soviet Air Force and Navy made the aircraft their standard medium bomber, also carrying mines and torpedoes. Over the vast distances of the Russian front, the German army depended heavily on the railways and the Soviet bomber regiments attacked them with great determination.

As with most Russian wartime aircraft, the Il-4 could absorb a substantial amount of damage and still fly, even when much of it was later made of wood. By the time production was terminated in 1944, over 6,800 had been rolled out and squadrons flew them until the surrender, though many were assigned to advanced training, reconnaissance and glider towing.

The Il-4 ended the war almost unknown to those outside the secretive Soviet Union but it was pivotal in eastern aerial strategy.

Specification (Il-4)

Powerplants: twin M-88B 1,100 hp 14-cylinder radials.
Dimensions: length 14.8 m (48 ft 6.5 in); height 4.2 m (13 ft 9 in); wing span 21.44 m (70 ft 4.25 in).
Weights: empty 6,000 kg (13,230 lb); operational 10,000 kg (22,046 lb).
Performance: maximum speed 410 km/h (255 mph); service ceiling 10,000 m (32,808 ft); range 2,600km (1,616 miles).
Armament: three to four 7.62 mm machine guns, one 12.7 mm gun; internal bomb load 1,000 kg (2,205 lb), external one 1,000 kg (2,205 lb) bomb or one to three torpedoes.

Lavochkin La-5/7

A La-7 of the 2nd Czechoslovak Fighter Regiment, serving with Soviet forces in 1945

Developed from the poorer performing LaGG-3 by refitting the aircraft with a more powerful radial engine, the La-5 (first flown as the LaGG-5 in January 1942) was an excellent fighter. Though it only carried two 20 mm cannon, the mostly wood La-5 was very light, exceptionally manoeuvrable and around 30 mph (48.28 km/h) faster than a Bf 109G at lower altitudes. The redesigned, lighter La-5FN was fitted with a more powerful engine and became the standard Lavochkin in the massive air battles of 1943, well able to meet the German opposition.

With a major redesign effort to reduce drag and
lower the aircraft's weight even more, another 20 mm
cannon was added along with increased underwing
bomb or rocket capacity and it was redesignated La-7.
Much to the delight of those who had flown earlier
models, the La-7 was just as manoeuvrable with a
higher top speed of 423 mph (680 km/h), equal to
anything on either side. In spite of having only two or
three cowl-mounted 20 mm cannon, all Lavochkins
had a more than sufficient punch, even in short
bursts.

With nothing to clutter the inside of the wing, the
design remained exceptionally clean, a major benefit
in air combat. Many of the Soviet Union's leading
aces, including top scorer Ivan Kozhedub (62 kills),
flew the Lavochkin fighters, finding them to be the
best in the air force inventory.

Specification: (La-5FN)

Powerplant: Shvetsov M-82FN 1,700 hp 14-cylinder radial.
Dimensions: length 8.46 m (27 ft 10.75 in); height 2.84 m (9 ft
3 in); wing span 9.8 m (32 ft 2 in).
Weights: empty unknown; operational 3,359 kg (7,406 lb).
Performance: maximum speed 650 km/h (403 mph); service
ceiling 10,000 m (32,800 ft); range 765 km (475 miles).
Armament: two 20 mm cannon, underwing bombs up to 150 kg
(330 lb) total.

Lavochkin-Gorbunov-Gudkov LaGG-3

The wooden LaGG-3 proved very robust, but its performance was inferior to most German fighters

Made primarily of wood, the LaGG-3 equalled the performance of most fighters designed in the late 1930s but sterling manoeuvrability was its strong point. As the RAF found out with the Mosquito, wood was not only very tough, able to absorb extensive punishment being pound for pound stronger than most metals, but it was easy to repair in the field.

Unfortunately, by the time Russia was in combat with first line German types on the Eastern Front, the LaGG-3's performance was substantially inferior. Pilots had to be quite sharp to hold their own, much less to prevail, over the enemy. With air force training and organisation thoroughly disrupted during the late 1930s by Stalin's purges, Russian pilots were seldom as well trained as their German opponents, and the reputation of both aircrew and aircraft was badly affected.

Though production lasted only from late 1940 (as the LaGG-1) to June 1942, at the beginning of 1942 more LaGGs were serving in the Soviet Air Force than any single fighter. The small fighter was flown by several units through to the end of the war.

Fortunately, the LaGG-3 could be adapted for quite a variety of armament depending on the mission, mounting anywhere from one 20 mm or 23 mm cannon and two machine guns to five machine guns and a cannon with underwing rockets or light bombs. A substantial improvement came when the airframe was adapted to a more powerful radial engine, leading to the outstanding La-5.

Specification (LaGG-3)

Powerplant: Klimov M-105P 1,050 hp 12-cylinder inline.
Dimensions: length 8.9 m (29 ft 1.25 in); height 3.22 m (9 ft 10 in); wing span 9.8 m (32 ft 2 in).
Weights: empty 2,630 kg (5,764 lb); operational 3,300 kg (7,257 lb).
Performance: maximum speed 560 km/h (348 mph); service ceiling 9,000 m (29,527 ft); range 650 km (404 miles).
Armament: one 20 mm cannon, two 12.7 mm machine guns, underwing rockets or bombs totalling 150 kg (330 lb).

Mikoyan-Gurevich MiG-3

Despite its relatively high speed, the MiG-3 proved disappointing as a fighter

Of the same generation as the LaGG-3, the MiG-3 was developed from the open cockpit MiG-1 as another wood fighter with an inline engine and markedly inferior armament. The long, heavy engine gave the fighter the appearance of 1930s racers along with a vicious swerve on take-off and landing.

After entering service at the same time production ceased, in late 1941, it was quickly proven to be inferior to German fighters since, unlike most Russian fighters, it was not very manoeuvrable. As a result the MiG, in spite of its respectable top speed of well over 400 mph (643 km/h), was withdrawn from the aerial superiority role and moved to reconnaissance and ground support as fast as it could be replaced.

Additional guns and wing racks were added but this only degraded the MiG's already marginal performance. The MiG-5 and high altitude MiG-7 were developed from the MiG-3 but very few were used in combat. By 1943 most MiG-3s had been withdrawn from combat entirely in favour of Yak-1s and La-5s.

German fighter pilots had little trouble downing most MiGs, which were considered easy meat, unless the Russian pilot pushed the throttle forward and had some altitude to extend away in a shallow dive. Many failed to escape and simply added to the enormous scores run up by the leading German aces flying on the Eastern front.

Specification (MiG-3)

Powerplant: Mikulin Am-35 1,200 hp 12-cylinder inline.
Dimensions: length 8.15 m (26 ft 9 in); height 2.61 m (8 ft 7 in); wing span 10.3 m (33 ft 9.59 in).
Weights: empty unknown; operational 3,490 kg (7,695 lb).
Performance: maximum speed 660 km/h (407 mph); service ceiling 12,000 m (39,370 ft); range 1,250 km (776 miles).
Armament: one 12.7 mm and two 7.62 machine guns with pods for two more 12.7 mm guns; six underwing rockets or two 100 kg (220 lb) bombs.

Petlyakov Pe-2

The tough Pe-2 served as a medium bomber, reconnaissance aircraft and long range fighter

One of the outstanding tactical attack aircraft of the war, regardless of nationality, the Petlyakov Pe-2 and Pe-3 design was in service from August 1940 until the surrender. Designed as a twin-engine fighter, the Pe-2 was a sleek aircraft from the start, giving German fighters fits when it could out run them, at times reaching over 400 mph (643 km/h). An excellent dive bomber, it could carry quite a respectable bomb load in both a bomb bay and under the wings. Later Pe-3bis versions reverted to the fighter role with forward firing 20 mm cannon and 12.7 mm machine guns. Many more (Pe-2Rs) were fitted as long-range reconnaissance platforms with an impressive array of cameras in the bomb bay.

As with most Russian aircraft, the Petlyakovs were rugged and could absorb a significant amount of battle damage. Crews, particularly the pilots, loved flying the aircraft, which often frustrated its fighter escort by cruising faster than they could keep up. With dive brakes out the aircraft was very stable in a dive and when the bombs were gone and the brakes were retracted, unlike the Stuka, Val and SBD, it would walk away from anything trying to shoot it down.

Specification (Pe-2)

Powerplants: twin Klimov M-105PF 1,260 hp 12-cylinder inlines.

Dimensions: length 12.66 m (41 ft 6 in); height 3.5 m (11 ft 6 in); wing span 17.2 m (56 ft 3.5 in).

Weights: empty 5,870 kg (12,943 lb); operational 7,700 kg (16,976 lb).

Performance: maximum speed 580 km/h (360 mph); service ceiling 8,800 m (28,870 ft); range 1,160 km (721 miles).

Armament: two 7.62 mm forward firing and two flexible 7.62 mm machine guns; 1,000 kg (2,205 lb) bomb load.

Polikarpov I-15/153

One of several I-15s captured by the Finns and flown against its former owners

Upon entering service in 1934, the I-15 was immediately one of the best biplane fighters of the 1930s, gaining an altitude record and immediate enthusiastic acceptance from its pilots. Over 800 were sent to fly with the Republicans in the Spanish Civil War, where they earned the nickname Chato (flat nose) during dive bombing and fighter cover missions. The last development of the basic airframe, the I-153, came along in 1939 with retractable landing gear and a more powerful engine, boosting top speed from 225 mph (362 km/h) to 265 mph (426 km/h). Several versions flew combat in China against the Japanese and then against Finland and Germany before Russia was invaded in June 1941. The Finns captured a number of I-15s in 1940 and put them back into service themselves.

Few fighters, regardless of performance, could outmanoeuvre an I-15 or 153. Even so, during the first week of combat against the Germans over Russia more than 2,200 of the small biplanes were lost. Contrary to what would seem to make sense, many I-153s were ordered to supplant monoplane I-16s to give pilots the benefit of extreme manoeuvrability as both aircraft were slower than most of the opposition. Several squadrons of Polikarpov biplanes served into 1942 to give close air support in spite of the danger involved, often becoming so much cannon fodder for eager Luftwaffe fighter pilots.

Specification (I-153)

Powerplant: Shvetsov M-63 1,000 hp 9-cylinder radial.
Dimensions: length 6.3 m (20 ft 3 in); height 2.9 m (9 ft 3 in); wing span 10 m (32 ft 9.75 in).
Weights: empty 1,450 kg (3,168 lb); operational 2,005 kg (4,431 lb).
Performance: maximum speed 430 km/h (267 mph); service ceiling 10,600 m (35,100 ft); range 480 km (289 miles).
Armament: two to four 7.62 mm machine guns; two 75 kg (165 lb) bombs or rockets.

Polikarpov I-16

The stubby I-16 was the most heavily armed fighter of the 1930s and performed well in Spain

When the tiny I-16 flew for the first time in December 1933, it was far ahead of any other fighter design in the world, featuring retractable landing gear, a cantilever wing and variable pitch propeller. When the Spanish Civil War broke out, almost 500 were put into service with the Republicans, surprising the enemy with outstanding manoeuvrability, firepower and rate of climb, leading to the opposition nickname Rata (Rat) and friendly name Mosca (Fly). The Soviet 20 mm cannon was the most powerful aircraft weapon in frontline service with any nation on the eve of World War II: it had a very high rate of fire and was extremely reliable. Another batch of I-16s was purchased by China to fight the Japanese, again surprising the other side with excellent performance.

When the Germans invaded Russia in June 1941, the I-16 was still Russia's most important fighter and, in spite of being obsolete, well over half of the 7,000 built were flown in action until 1943. One of the most startling uses of the tiny but rugged fighter came in ramming attacks. Pilots were taught to hit the tail surfaces of German bombers, then bail out. In theory, the strength of the I-16 would allow the pilot grace to bail out afterwards.

If German pilots decided to outmanoeuvre the I-16 in dogfights, which invariably bleed off speed, they were usually caught by surprise as the Russian pilot quickly got the upper hand. However, against slashing climbing/diving attacks, the I-16 was in trouble.

Specification (I-16 Type 18)

Powerplant: Shvetsov M-62R 1,000 hp 9-cylinder radial.
Dimensions: length 6.13 m (20 ft 1.25 in); 2.57 m (height 8 ft 5 in); wing span 9.18 m (29 ft 6.5 in).
Weights: empty 1,412 kg (3,110 lb); operational 1,831 kg (4,034 lb).
Performance: maximum speed 463 km/h (288 mph); service ceiling 8,998 m (29,500 ft); range 805 km (500 miles).
Armament: two 7.62 mm machine guns, two 20 mm cannon.

Tupolev SB-2

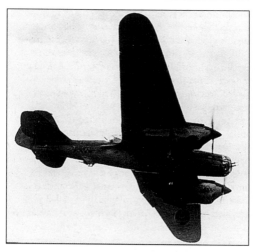

The SB-2 was one of the best medium bombers in service at the beginning of World War II

In spite of being developed in the early 1930s, from the time it entered Russian service in early 1936 until final production delivery in 1942, the SB-2 was superior to most early wartime allied bombers of the same size. Soviet bomber regiments equipped with SB-2s took part in the invasion of Finland in 1940.

Given to the Republicans in the Spanish Civil War and the Chinese to stem the Japanese invasion, the aircraft proved excellent in low-level and medium altitude attack with a top speed higher than most fighters. Pilots found that even as it gained weight through successive versions, increased engine horsepower and streamlining yielded enough performance to compensate.

As the I-16 was the USSR's main fighter when war broke out in June 1941, the SB-2 was its main bomber, flying through those horrendous first two years of combat. As Russia's first stressed metal skin bomber, the aircraft had enough advanced (for its time) features to keep it up to date. Nevertheless, as more modern types began to appear, the SB-2 was relegated to safer roles, particularly night bombing.

Specification (SB-2bis)

Powerplants: twin M-103 1,100 hp 12-cylinder inlines.
Dimensions: length 12.27 m (40 ft 3.25 in); height 3.25 m (10 ft 8 in); wing span 20.35 m (66 ft 8.5 in).
Weights: empty 4,903 kg (10,800 lb); operational 7,806 kg (17,196 lb).
Performance: maximum speed 450 km/h (280 mph); service ceiling 10,675 m (35,000 ft); range 1,599 km (994 miles).
Armament: four flexible 7.62 mm machine guns; internal bomb load up to 600 kg (1,320 lb).

Tupolev Tu-2

The Tupolev Tu-2 continued in service after 1945, some taking part in the Korean war

First flown in October 1940, the Tu-2 was an outstanding attack aircraft from the time it entered service in August 1942. Built in far fewer numbers than its older Pe-2 stablemate, the Tupolev was larger and more effective, able to carry quite a powerful array of armament and bombs in a number of roles, including dive bombing, close air support, medium bomber and attack fighter. By 1943 the Soviet forces were proving more capable of integrating airpower with their ground operations. However, bomber units were often rigidly assigned to discrete Fronts (Army Groups) which made them less flexible than their German opponents.

Crews were universally happy with their machines. Pilots could manoeuvre it like a fighter and it was fast and rugged enough to survive.

Through to the end of World War II the basic design remained unchanged, a testimony to its original integrity and adaptability. Tu-2 squadrons were deployed across Russia during the war, taking part in every major battle, pushing the Germans back with a tenacity linked to a large degree with the aircraft's rugged nature. So useful was the Tupolev that it remained in Soviet service through to the end of the decade, flew in the Korean War with North Korea and into the 1960s with the People's Republic of China and other Communist oriented countries.

Specification (Tu-2)

Powerplants: twin Shvetsov ASh-82FN 1,850 hp 14-cylinder radials.

Dimensions: length 13.82 m (45 ft 3.75 in); 4.21 m (height 13 ft 9.5 in); wing span 18.87 m (61 ft 10.5 in).

Weights: empty 8,280 kg (18,240 lb); operational 12,811 kg (28,219 lb).

Performance: maximum speed 550 km/h (342 mph); service ceiling 9,506 m (31,168 ft); range 2,499 km (1,553 miles).

Armament: three flexible 12.7 mm machine guns, two forward firing 20 mm cannon; internal bomb load up to 2,270 kg (5,000 lb).

161

Yakovlev Yak-1/3/7/9

One of the Yak-9DDs of the Balkan air force that escorted US bombers over Yugoslavia

This enormously successful series of fighters began with the Yak-1's first flight in March 1939 and stretched all the way to the Yak-9P's service in the Korean War over a decade later. The wood-winged, steel tube fuselage fighter was an immediate success, leading to the Yak-7B, which entered production in early 1942.

A further refinement of the airframe led to the Yak-3 (at first Yak-1M), entering service in July 1943 with a bubble style canopy and a number of design improvements, yielding an outstanding top speed of 447 mph (719 km/h) in later versions. The Germans recognized it as one of the best fighters they opposed. Only the patchy quality of Soviet fighter pilots prevented it achieving greater results.

The Yak-7 turned into the Yak-9, which had metal wing spars and became the most numerous of all versions. Flying in a number of roles, including close air support and long-range escort fighter, the Yak-9 was easily superior to the Messerschmitt Bf 109G and the Yak-9U was better than both the 109 and the 190. The Free French Normandie-Niemen squadron, formed from volunteers after the surrender of France in 1940 flew Yak-9s, then switched to the higher performing Yak-3s. The French achieved the impressive total of 273 kills over the Germans. By the end of the production run, over 37,000 Yak-1/3/7/9s had been delivered, second in number only to the Illyushin Il-2.

Specification (Yak-3)

Powerplant: Klimov VK-105PF-2 1,225 hp 12-cylinder inline.
Dimensions: length 8.5 m (27 ft 10.25 in); height 2.39 m (7 ft 10 in); wing span 9.21 m (30 ft 2.25 in).
Weights: empty 2,252 kg (4,960 lb); operational 2,662 kg (5,864 lb).
Performance: maximum speed 650 km/h (404 mph); service ceiling 10,812 m (35,450 ft); range 814 km (506 miles).
Armament: two 12.7 mm machine guns, one 20 mm cannon.

Armstrong Whitworth Whitley

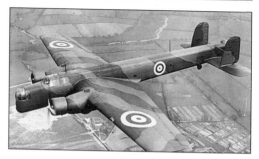

Whitleys flew on the early RAF night raids against Germany during 1941-2

When the prototype Whitley flew in March 1936 its modern metal construction was a preview of things to come, yet its unrefined aerodynamics doomed it to mediocrity. Even so, the aircraft had the distinction of being in the forefront of RAF Bomber Command's first efforts against the enemy, on 19 March 1940 dropping the first bombs on Germany since the end of World War I.

Until the newer types could come along, the Whitley continued to press into enemy airspace at night through to early 1943, suffering severely but following the mandate to serve in the only offensive striking force available to the Allies at the time.

Well over 100 were modified for Coastal Command patrol work, successfully seeking out U-boats with radar and hitting them as often as possible. This was a vital task which, if allocated more aircraft could have saved many Allied lives in the mid-Atlantic.

As it was withdrawn from bomber squadrons, the Whitley served to train aircrews, familiarize parachute troops with combat drops, field new radars and test electronics countermeasures. Two Whitley squadrons flew supplies to the French resistance.

The last Whitley rolled out of the factory in June 1943, probably later than most had intended. Crews were not overly fond of the 'Whitbox' as it was slow, sluggish on the controls and poorly equipped to survive against cannon-armed fighters.

Specification (Whitley V)

Powerplants: twin Rolls-Royce Merlin X 1,145 hp 12-cylinder inlines.

Dimensions: length 21.5 m (70 ft 6 in); 4.57 m (height 15 ft); wing span 25.6 m (84 ft).

Weights: empty 8,762 kg (19,330 lb); operational 15,195 kg (33,500 lb).

Performance: maximum speed 367 km/h (228 mph); service ceiling 5,368 m (17,600 ft); range 2,655 km (1,650 miles).

Armament: one .303 calibre machine gun in nose turret, four .303 guns in tail turret; internal bomb load 3,178 kg (7,000 lb).

Avro Lancaster

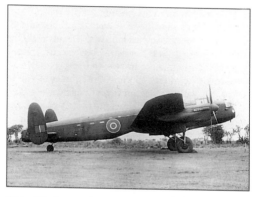

One of the famous Lancasters of 617 squadron, prepared for the 'dam busting' raid of May 1943

The most famous and respected of British bombers, the Lancaster was born from the failed twin-Vulture-engined Manchester. When the initial prototype Lanc first flew on 9 January 1941, with four Rolls-Royce Merlin engines, the magic was immediate. One of the Manchester's fortunate bequests was an enormous bomb bay designed to carry massive 4,000-pound (1816 kg) bombs. This resulted in the Lancaster being the only wartime bomber capable of hauling the giant 22,000-pound (9988 kg) Grand Slam.

The type's first combat mission was flown on 2 March 1942, but fame was guaranteed when No.617 Squadron carried out the Dam Busters raid of 17/18 May 1943. Pilots and crews were unanimous in their praise of the large bomber. Not only was it manoeuvrable enough to 'corkscrew' dive away from pursuing night fighters, but it could take quite a bit of punishment. Defensive armament was among the most effective fitted to any British wartime aircraft. The Lanc quickly became the most capable of the RAF's bomber aircraft, heading toward Germany on as many nights as possible through to the end of the war. Streams of Lancasters stretching for hundreds of miles flew the devastating Dresden missions in February 1945, laying waste to the city.

Specification (Lancaster I)

Powerplants: four Rolls-Royce Merlin 20 1,280 hp 12-cylinder inlines.

Dimensions: length 21.15 m (69 ft 4 in); height 5.97 m (19 ft 7 in); wing span 31.09 m (102 ft).

Weights: empty 16,738 kg (36,900 lb); operational 24,062 kg (53,000 lb).

Performance: maximum speed 462 km/h (287 mph); service ceiling 7,470 m (24,500 ft); range 2,670 km (1,660 miles).

Armament: two .303 calibre machine guns each in nose and top turrets, four .303 guns in tail turret; normal bomb load 6,356 kg (14,000 lb), maximum one 9,988 kg (22,000 lb) bomb.

Bristol Beaufighter

SR919, the ultimate Beaufighter/torpedo-bomber carries radar in its nose and underwing rocket rails

The Bristol company's idea of a heavily armed, long-range fighter, the Beaufighter was a rapid modification of the Beaufort, using the bomber's wing, tail, landing gear and basic internal equipment. The prototype flew only six months later, in July 1939, placing the much needed fighter into service by July 1940. Though it was quite a manoeuvrable powerhouse, particularly as a night fighter and anti-shipping strike aircraft, the Beau was vicious with one engine out and had a marked propensity to swerve on take-off and landing.

In spite of being slower than most had hoped, never seeming to get the increased power engines it needed, the rugged radar-equipped night fighter was one of the major reasons for the Luftwaffe's failure during the night Blitz on England in 1940 and 1941. When RAF Coastal Command got their Beaus they hung everything possible from them, including torpedoes and rockets, making enormous dents in German shipping. Built in Australia as well, Beaufighters served across the Far East and the Pacific, primarily as strike aircraft. Pilots, though wary of going in if one engine failed, particularly on take-off, were loud in their praise of the general handling and stout nature of their aircraft, which seemed to be in just the right place at the right time.

Specification (Beaufighter X)

Powerplants: twin Bristol Hercules XVII 1,770 hp 14-cylinder radials.

Dimensions: length 12.7 m (41 ft 8 in); height 4.83 m (15 ft 10 in); wing span 17.63 m (57 ft 10 in).

Weights: empty 7,082 kg (15,600 lb); operational 11,531 kg (25,400 lb).

Performance: maximum speed 502 km/h (312 mph); service ceiling 8,083 m (26,500 ft); range 2,478 km (1,540 miles).

Armament: four 20 mm cannon (plus six .303 machine guns in fighter), one flexible .303 gun in dorsal position; bomb load 908 kg (2,000 lb), eight rockets or one torpedo.

Bristol Beaufort

This Beaufort VIII of 100 squadron, RAAF carried out a record 145 sorties over the SW Pacific

A beefy offshoot of the Blenheim, the Beaufort first flew in October 1938 with deliveries to the RAF beginning a year later. During initial design efforts, the Australians signed an agreement to build the bomber, leading to their largest wartime aircraft manufacturing effort of the war. Among other major modifications they made, instead of English powerplants they installed Pratt & Whitney R-1830s, easier to maintain and to procure. RAAF Beauforts, almost unknown in the shadow of more famous Allied aircraft, saw action across the Pacific.

After experiencing some teething problems with the Taurus engines, which tended to overheat, the first RAF Coastal Command Beaufort squadrons, carrying torpedoes as their primary armament, entered action with a vengeance, sinking a number of German ships. These anti-shipping missions, later flown by Beaufighters, were among the most dangerous of the war, requiring pilots to fly a steady course, extremely low to deliver their ordnance in the face of intense anti-aircraft fire. Skip bombing became a standard tactic, a principle later used to attack the Ruhr Valley dams. Almost unheralded, Coastal Command Beaufort crews were a major striking force during the first few years of the war.

Specification (Beaufort I)

Powerplants: twin Bristol Taurus VI 1,130 hp 14-cylinder radials.

Dimensions: length 13.47 m (44 ft 2 in); height 4.35 m (14 ft 3 in); wing span 17.63 m (57 ft 10 in).

Weights: empty 5,945 kg (13,107 lb); operational 9,630 kg (21,230 lb).

Performance: maximum speed 418 km/h (260 mph); service ceiling 5,030 m (16,500 ft); range 2,574 km (1,600 miles).

Armament: two .303 calibre machine guns in top turret, one remotely controlled .303 gun in nose with option of four forward firing .303 guns; internal bomb load 908 kg (2,000 lb) or one partially external torpedo.

Bristol Blenheim

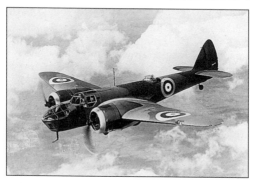

A Blenheim IV light bomber: a successful design widely exported during the 1930s

Developed from a fast executive transport, the Blenheim prototype flew for the first time in April 1935 as the nation's first all-metal, stressed skin, retractable landing gear, controllable pitch propeller monoplane. With an astonishing (for then) speed of over 200 mph (320 km/h), the aircraft was ordered in very large numbers despite the economic depression. By the time World War II began, the Blenheim was serving RAF Bomber Command and was soon hitting targets across the Continent and in Germany. The bomber was far enough ahead of its time to be licence-built in Canada (as the Bolingbroke) and sold to Finland, France, Greece, Yugoslavia, Lithuania, Portugal, Romania and Turkey.

Many Blenheims were converted to night fighters and flown with the world's first airborne intercept radar, pioneering what was to become a significant area of air combat. The Mk.IV, with an asymmetric stepped cockpit and nose, became the main combat version, bearing the brunt of the RAF's valiant tries at daylight strike. On the whole, crews enjoyed their aircraft but, as so many found out, flying unescorted daylight bombing missions was one of the most dangerous things to do in World War II with high casualty rates. That the Blenheim did its job so well in the face of such opposition speaks a great deal for a mid 1930s design.

Specification (Blenheim IV)

Powerplants: twin Bristol Mercury XX 920 hp 9-cylinder radial.
Dimensions: length 13.04 m (42 ft 9 in); height 3.91 m (12 ft 10 in); wing span 17.17 m (56 ft 4 in).
Weights: empty 4,445 kg (9,790 lb); operational 6,537 kg (14,400 lb).
Performance: maximum speed 428 km/h (266 mph); service ceiling 9,607 m (31,500 ft); range 3,138 km (1,950 miles).
Armament: two .303 calibre machine guns each in chin and top turrets, one .303 gun in nose; internal bomb load 454 kg (1,000 lb).

De Havilland Mosquito

Mosquito FIIs defended the UK and flew night intruder missions over Germany

Universally loved by its crews, hated and copied by its enemies, the 'Mossie' was born beautiful as a high-speed bomber built mostly of molded plywood, a nonstrategic material. When Geoffrey de Havilland, Jr. flew the prototype on 25 November 1940, he knew his father's company had created something unique. Mosquitoes became some of the finest combat types of the war, serving equally well in multiple roles, including night fighter and anti-shipping strike. When the first F.II night fighter flew in May 1941 planners knew they had one of the most potent of all night weapons, a faith well proven throughout the war as crews racked up the majority of RAF night kills.

An obsession with speed built in at the expense of defensive armament gave the type the lowest Bomber Command loss rate of the war, something very much appreciated in the bloody business of night bombing. The first Mk.IV bomber versions entered combat with No.105 Squadron in May 1942. The aircraft was particularly well suited for pathfinding, preceding the bomber streams to drop target markers, then outrun enemy night fighters on the way out. Until 1944, Mossies were the fastest RAF aircraft in service and in Bomber Command they remained the fastest type until introduction of the jet English Electric Canberra in 1951. The pressurized photo-recce version, packed with cameras and extra fuel, could cruise over enemy territory at over 40,000 feet (12,200 m) in excess of 400 mph (644 km/h) if necessary.

Specification (Mosquito IV)

Powerplants: twin Rolls-Royce Merlin 21 1,230 hp 12-cylinder inlines.

Dimensions: length 12.35 m (40 ft 6 in); height 3.81 m (12 ft 6 in); wing span 16.51 m (54 ft 2 in).

Weights: empty 6,401 kg (14,100 lb); operational 10,215 kg (22,500 lb).

Performance: maximum speed 611 km/h (380 mph); service ceiling 10,522 m (34,500 ft); range 2,993 km (1,860 miles).

Armament: internal bomb load 908 kg (2,000 lb) to 1,816 kg (4,000 lb).

Fairey Battle

Battles suffered terrible losses in France: K9353 was one of them, posted missing on 13 May 1940

When the Battle first flew in March 1936 it appeared to be a vision of the future, sleek with broad wings and the new Merlin engine. As a result, the type was bought in large numbers as a strike bomber, equipping a large proportion of RAF bomber squadrons from June 1937.

By the time World War II started over 1,000 were in first line service and just a few days into the war ten squadrons were flown to France as Britain's most formidable answer to the German onslaught. During the 'Phoney War' of the next eight months the Battle fared well since there were so few enemies to deal with, but by May 1940 it was being slaughtered in almost record numbers with losses of well over 50% per mission. The first RAF VC of the war was awarded (posthumously) to the leader of a Battle formation, attacking the Albert Canal bridges despite the suicidal odds.

Not only was it no longer faster than most fighters, but it was terribly vulnerable to both anti-aircraft fire and fighters with its single defensive .303 machine gun. By the end of the year the Battle had been withdrawn from combat service and the surviving examples were sent to Canada, Australia, Rhodesia and South Africa as advanced trainers with dual controls as well as target tow aircraft. For such prewar promise, the Battle was one of the most disappointing of all RAF aircraft.

Specification (Battle II)

Powerplant: Rolls-Royce Merlin II 1,030 hp 12-cylinder inline.
Dimensions: length 12.85 m (42 ft 1.75 in); height 4.72 m (15 ft 6 in); wing span 16.46 m (54 ft).
Weights: empty 3,015 kg (6,647 lb); operational 4,895 kg (10,792 lb).
Performance: maximum speed 388 km/h (241 mph); service ceiling 7,620 m (25,000 ft); range 1,448 km (900 miles).
Armament: one forward firing .303 calibre machine gun, one flexible rearward firing .303 gun; internal bomb load 454 kg (1,000 lb).

Fairey Firefly

The Firefly was the best British carrier aircraft of the war

Patterned after the earlier Fulmar, the Firefly turned out to be Britain's finest World War II carrier aircraft. Though not flying its first combat mission with the Royal Navy until July 1944, against the Tirpitz in Norway, it was rushed into every theatre of war as fighter, night fighter, fighter-bomber and reconnaissance platform. This well-muscled Rolls-Royce Griffon-powered aircraft became one of the first modern multirole combat aircraft, taking on numerous jobs once assigned to several different aircraft.

The first versions of the aircraft had the radiator in the nose, typical of so many British liquid-cooled engine aircraft, but later versions, from the Mk.IV, had the heat exchangers buried in the wings, giving a very pleasing, streamlined effect. From the start the armament of four 20 mm cannon and a 2,000 pound (908 kg) external bomb load was excellent. The flaps enabled the aircraft to be flown at very low speeds, making it a delight and quite safe, even when shot up, to bring back aboard ship. With search radar the Firefly became a potent hunter-killer for the fleet and several were turned into night fighters.

Specification (Firefly I)

Powerplant: Rolls-Royce Griffon XII 1,990 hp 12-cylinder inline.

Dimensions: length 11.46 m (37 ft 7 in); height 4.14 m (13 ft 7 in); wing span 13.56 m (44 ft 6 in).

Weights: empty 4,423 kg (9,750 lb); operational 6,359 kg (14,020 lb).

Performance: maximum speed 509 km/h (316 mph); service ceiling 8,535 m (28,000 ft); range 933 km (580 miles).

Armament: four 20 mm cannon; external bomb load 908 kg (2,000 lb).

Fairey Swordfish

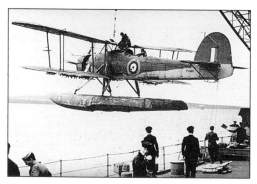

A Fairey Swordfish Mk I being launched from the battleship HMS Malaya

An unabashed antique by World War II, having made its first flight in 1933, the Swordfish broke all convention to become one of the most effective seaborne combat aircraft of World War II. When other powers were flying monoplane, metal stressed skin scout torpedo bombers, for the first two years of the war the Royal Navy's striking power was based almost entirely on the biplane, wood and fabric Swordfish, nicknamed 'Stringbag' because of its wires and bracing. Floatplane versions served aboard battleships and cruisers; Swordfishs with conventional undercarriages flew from British aircraft carriers. The lumbering Swordfish quickly scored a number of triumphs against German ships, neutralized the Vichy French fleet at Oran and sank the Italian fleet at Taranto in November 1940, the first major carrier strike of the war. The feat at Taranto proved to the Japanese that a carrier force could attack and destroy the American fleet at Pearl Harbor. In May 1941 Swordfish took part in the sinking of the mighty German battleship Bismarck by disabling its rudders, making it vulnerable to the British battle force. When the 'Stringbag' was finally retired later in the war it had been effective far beyond its years, due in large measure to excellent pilots and clever tactics, proving carrier aviation was a war-winning capability.

Specification (Swordfish I)

Powerplant: Bristol Pegasus IIIM3 775 hp 9-cylinder radial.

Dimensions: length 11.08 m (36 ft 4 in); height 4.11 m (13 ft 5.75 in); wing span 13.87 m (45 ft 6 in).

Weights: empty 2,134 kg (4,700 lb); operational 3,677 kg (8,100 lb).

Performance: maximum speed 224 km/h (139 mph) (without torpedo); service ceiling 3,660 m (12,000 ft); range 879 km (546 miles).

Armament: one forward firing .303 calibre machine gun, one flexible rearward firing .303 gun; bomb load 681 kg (1,500 lb) or one torpedo.

Gloster Gladiator

This Gladiator Mk II served as part of flight deployed to the Shetlands in 1939

Another of World War II's strange biplane antiques, the Gladiator lasted far longer than it should have after being ordered in the summer of 1935. So popular was the aircraft, it was bought by 13 countries in addition the RAF and the Royal Navy (as the Sea Gladiator). In spite of its configuration, it was fairly clean by the standards of the day with four machine guns (heavier than normal), an enclosed cockpit and landing flaps. Against great odds, Gladiators defended Britain from early German bombing raids, flew from Norway off frozen lakes when their carriers were hit and only three, the famous Faith, Hope and Charity, defended Malta day after day in June 1940.

As with most biplane fighters, the Gladiator was a pilot's aircraft, manoeuvrable and responsive to the slightest input. This sterling quality often made up for lack of speed in fighting superior German machines but pilots had to know how to use it or get shot down. Since the basic airframe was covered with fabric, it was also easy to put back into service when shot up by doping fabric patches onto the holes. The Gladiator was all that was available in some places and it held on until the more modern types could arrive.

Specification (Gladiator I)

Powerplant: Bristol Mercury IX 840 hp 9-cylinder radial.
Dimensions: length 8.36 m (27 ft 5 in); height 3.15 m (10 ft 4 in); wing span 9.83 m (32 ft 3 in).
Weights: empty 1,566 kg (3,450 lb); operational 2,156 kg (4,750 lb).
Performance: maximum speed 407 km/h (253 mph); service ceiling 10,065 m (33,000 ft); range 708 km (440 miles).
Armament: four .303 calibre machine guns.

Gloster Meteor

The RAF's first operational jet fighter, the Meteor was deployed to Holland in April 1945

The Gloster Meteor Mk.I was the first jet aircraft to enter service with the RAF and the only Allied jet aircraft to see action in World War II. With many of the same development problems experienced in Germany, the Meteor flew for the first time, in spite of a short, tentative hop in July 1942, on 5 March 1943. With many excellent propeller-driven types winning the war, the Gloster jet became an aircraft in search of a mission. When the first V-1 flying bomb entered British airspace on June 13, 1944 the mission was suddenly evident. By the end of the month, more than 2,400 V-1s had been launched at London, one third of them reaching the city.

Though the Meteor Mk.I was barely equal in speed to most piston fighters at low level, that was just enough to turn the jet into a flying-bomb killer at the hands of No.616 Squadron, the only unit to fly the type in action, beginning on 27 July 1944. With the Mk.III version and new Rolls-Royce 2,000-pound-thrust engines came a leap in performance and a top speed of 495 mph (796 km/h) at 30,000 feet (9150 m). Moving to the Continent in early 1945 on air defence, 616 Squadron was based near Nijmegen by April. After the war, the Meteor became the RAF's primary jet aircraft, serving until the 1960s.

Specification (Meteor I)

Powerplants: twin Rolls-Royce Welland W.2B/23 1,700 lb) thrust centrifugal flow turbojets.
Dimensions: length 12.61 m (41 ft 4 in); height 3.96 m (13 ft); wing span 13.11 m (43ft).
Weights: empty 3,695 kg (8,140 lb); operational 6, 265 kg (13,800 lb).
Performance: maximum speed 660 km/h (410 mph); service ceiling 12,200 m (40,000 ft); range 1,609 km (1,000 miles).
Armament: four 20 mm cannon.

Handley Page Hampden

Hampdens took part in the first night raids on Germany after daylight attacks proved too costly

Another of the RAF's mid 1930s fast bomber developments, the Hampden entered service just before the war broke out. However, unlike the Blenheim I, its stablemate, it had four times the payload and twice the range. As a result, Hampdens were deep into enemy territory from the very beginning of Bomber Command's offensive. Very manoeuvrable and, as a result, more likely to survive, the bomber was a very important part of early war bombing strategy. As other more capable aircraft took its place, the 'Flying Suitcase' laid mines and carried torpedoes against shipping as far as Russia.

As happened with the other RAF bombers, when enemy fighters began to shoot down alarming numbers the offensive was switched to night. Though this did have the initial result of lowering losses, eventually the Germans developed an outstanding night defence capability and more crews were killed per sortie at night than during the day. Nevertheless, Hampden crews were often farther into the Third Reich than any other British type, hitting the enemy when there was nothing else on which to base an offensive.

Specification (Hampden 1)

Powerplants: twin Bristol Pegasus XVIII 1,000 hp 9-cylinder radials.

Dimensions: length 16.33 m (53 ft 7 in); height 4.37 m (14 ft 4 in); wing span 21.08 m (69 ft 2 in).

Weights: empty 5,343 kg (11,780 lb); operational 8,508 kg (18,756 lb).

Performance: maximum speed 409 km/h (254 mph); service ceiling 5,790 m (19,000 ft); range 1,762 km (1,095 miles).

Armament: two flexible rear firing .303 calibre machine guns each in top and bottom positions, one fixed and one flexible forward firing .303 gun; internal bomb load 1,816 kg (4,000 lb) of bombs, mines or one torpedo, external racks for two 227 kg (500 lb) bombs.

Handley Page Halifax

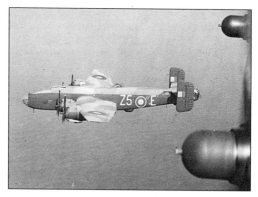

This Halifax was lost in 1944, colliding with a Lancaster returning from a raid against Essen

The second British four-engine bomber to enter combat behind the Stirling, the Handley Page Halifax flew its first mission with No.35 Squadron in March 1941. In spite of a number of teething problems, the bomber went through several marks to become very successful at its job, unfortunately overshadowed by its stablemate, the Lancaster, as the B-24 was overshadowed by the B-17. The 'Halibag' was much loved by its crews and was used in a greater variety of jobs than most heavy bombers. Gross weight climbed from just over 25,000 lb (11,350 kg) to over 60,000 lb (27,240 kg) as the aircraft went through various versions, all fitted with the enormous bomb bay featured in most British heavy bombers.

Changing shape slowly, the nose, vertical tail surfaces, wings and engines all metamorphosed, the major change being a switch from liquid-cooled inline Merlin to air-cooled radial Hercules engines. The ultimate Mk.VI was a jack of all trades and master of many as a bomber, Coastal Command patrol aircraft, sub hunter and weather ship. Several were converted to transports to drop paratroops, tow gliders and serve as electronics countermeasures platforms. Not until the early 1950s were the last Halifaxes retired from service.

Specification (Halifax III)

Powerplants: four Bristol Hercules XVI 1,615 hp 14-cylinder radials.

Dimensions: length 21.36 m (70 ft 1 in); height 6.32 m (20 ft 9 in); wing span 104 ft 2 in (31.75 m)

Weights: empty 17,345 kg (38,240 lb); operational 24,697 kg (54,400 lb) (29,510 kg (65,000lb) overload).

Performance: maximum speed 454 km/h (282 mph); service ceiling 7,315 m (24,000 ft); range 1,658 km (1,030 miles).

Armament: four .303 calibre machine guns each in top and rear turrets, one flexible .303 gun in nose; bomb load 5,902 kg (13,000 lb)

Hawker Hurricane

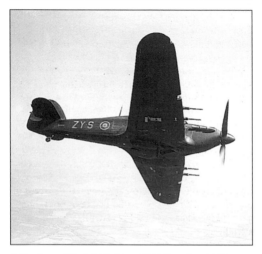

A Hurricane IIC night fighter of 247 Squadron RAF

First flown in November 1935, the Hurricane was
built in large enough numbers and was effective
enough to become the main reason why the RAF won
the Battle of Britain in the summer of 1940. Though
primitive in many ways by the time the war began,
the 'Hurri' was easy to fly, stable as a gun platform,
carried eight .303 guns and rugged enough to

withstand extensive battle damage. Most of RAF Fighter Command's kills during the Battle were made by Hurricanes and it could out turn any other modern fighter in the sky.

When newer types began to replace it as a first line fighter, the aircraft was turned into a fighter-bomber and ground support aircraft par excellence with four 20 mm or two 40 mm cannon, fighting through the Western Desert and the Mediterranean with distinction. Sea Hurricanes were crucial in the defence of Malta during the summer of 1942 and around 3,000 were given to Russia. Eighteen other countries, in addition to Britain, received Hurricanes and put them to good use. When the last one was delivered in September 1944 almost 13,000 had been built, quite a record for what was by 1941, essentially, an obsolete aircraft.

Specification (Hurricane I)

Powerplant: Rolls-Royce Merlin II 1,030 hp 12-cylinder inline.
Dimensions: length 9.58 m (31 ft 5 in); height 3.99 m (13 ft 1 in); wing span 12.19 m (40 ft).
Weights: empty 2,120 kg (4,670 lb); operational 2,996 kg (6,600 lb).
Performance: maximum speed 512 km/h (318 mph); service ceiling 10,980 m (36,000 ft); range 740 km (460 miles).
Armament: eight .303 calibre machine guns.

Hawker Typhoon

King George VI inspecting Typhoons of 183/164 Squadrons before their deployment to Normandy

Encouraged by the potential of the Napier Sabre 2,000+ hp flat-H engine, Hawker Aircraft's Sydney Camm submitted a proposal to the British Air Ministry in April 1937 for a new fighter, eventually named the Typhoon, designed around this radical, sleeve-valve powerplant. The prototype flew for the first time on February 24, 1940, but the Sabre was constantly troublesome. The Hawker fighter was an airplane looking for a mission since it had very poor performance above 20,000 feet (6,100 m), due in large measure to its typical thick Camm airfoil.

In spite of constant engine changes and a number of irritating quirks, including a maddening tendency not to start in cold weather, the Typhoon Mk.IA, with twelve .303 machine guns, was thrust into service with No.56 Squadron in late 1941. The Mk.IB 'Tiffy' was given four 20 mm Hispano cannon in place of the .303s and a clear rear vision Perspex panel in place of the metal-panelled rear fairing, though the problem pilots were having getting a good look aft was not solved until a bubble canopy was fitted in mid 1943. Many of the aircraft's problems were ironed out, resulting in an excellent ground attack fighter which could carry rockets and bombs in addition to the very effective cannon.

Specification (Typhoon IB)

Powerplant: Napier Sabre II 2,180 hp 24-cylinder flat-H inline.
Dimensions: length 9.73 m (31 ft 11 in); height 4.66 m (15 ft 3.5 in); wing span 12.67 m (41 ft 7 in).
Weights: empty 3,992 kg (8,800 lb); operational 6,015 kg (13,250 lb).
Performance: maximum speed 663 km/h (412 mph); service ceiling 10,736 m (35,200 ft); range 820 km (510 miles).
Armament: four 20 mm cannon; external bomb load up to 908 kg (2,000 lb) of bombs or eight rockets.

Hawker Tempest

Tempests led the battle against V1 flying bombs

When Sydney Camm initiated a number of studies to improve the Typhoon's high-altitude performance, reluctantly he created a new, thinner wing, lengthened the forward fuselage 22 inches for an additional fuel tank and, ultimately, added a dorsal fin sorely needed for correction of directional instability. Though he had designed wing radiators to replace the deep chin, estimated to reduce drag by two thirds, wartime expediency pressed him to stick with the original chin configuration.

The operational fighter, with lighter and shorter 20 mm cannon, was rated at 426 mph (685 km/h) at 18,500 feet (5,642 m). When the newest Hawkers went operational in April 1944, timing to counter the flying bomb assault was nearly perfect. Pilots quickly came to appreciate their Tempests, racking up an outstanding kill record, particularly against V-1s. Those who remember say there was nothing quite like a 24-cylinder Napier Sabre II at full song. When the Centaurus radial engine was finally perfected for installation in the Tempest Mk.II, the results were far better. Not only did top speed increase but it was quieter and better on the controls, but it did not enter service until after the war.

Specification (Tempest V)

Powerplant: Napier Sabre II 2,180 hp 24-cylinder flat-H inline.
Dimensions: length 10.26 m (33 ft 8 in); height 4.90 m (16 ft 1 in); wing span 12.5 m (41 ft).
Weights: empty 4,131 kg (9,100 lb); operational 6,129 kg (13,500 lb).
Performance: maximum speed 687 km/h (427 mph); service ceiling 11,285 m (37,000 ft); range 1,190 km (740 miles).
Armament: four 20 mm cannon; external bomb load 908 kg (2,000 lb) of bombs or eight rockets.

Short Sunderland

The magnificent Sunderland proved so tough, German pilots dubbed it 'the flying porcupine'

The Sunderland, in spite of being one of the last flying boats designed, was durable enough to fly in service from June 1938 until the early 1960s. When World War II started, Coastal Command's three Sunderland squadrons were immediately pressed into ocean patrol, search and rescue, and, in January 1940, made the first of many U-boat kills. With their great endurance, Sunderlands spotted German ship movements when other types were forced back to base due to lack of fuel. The aircraft's excellent defensive armament became famous to the Germans, who nicknamed the aircraft 'Flying Porcupine.'

The large flying boat was in demand for convoy escort throughout the war due not only to its striking power but its ability to land on the water for immediate rescue. During many desperate evacuation operations early in the war Sunderlands could carry large numbers of personnel in an almost continuous stream without the need for land runways. As more capability was added to the airframe, anti-shipping strike was undertaken across the globe. Often overshadowed by more glamorous aircraft, the Sunderland ended the war as one of the most effective weapons employed by any side and it was the only RAF aircraft to be used from the beginning to the end of the Korean War.

Specification (Sunderland V)

Powerplants: four Pratt & Whitney R-1830 Twin Wasp 1,200 hp 14-cylinder radials.

Dimensions: length 26.03 m (85 ft 4 in); height 10.03 m (32 ft 10.5 in); wing span 34.38 m (112 ft 9.5 in).

Weights: empty 16,798 kg (37,000 lb); operational 27,240 kg (60,000 lb).

Performance: maximum speed 343 km/h (213 mph); service ceiling 5,455 m (17,900 ft); range 4,795 km (2,980 miles).

Armament: four each .303 calibre machine guns in nose and tail turrets, two .303 guns in top turret, two flexible .50 calibre guns in beam positions; bomb load 908 kg (2,000 lb).

Short Stirling

The Mk III was the last bomber variant; later Stirlings served as glider tugs or transports

The Short Stirling was the first four-engine monoplane bomber to enter service with the RAF and the first to see action in World War II. Unfortunately, it was underpowered, but then again what would not be at a service weight of 31 tons in 1939. A mere 4,000 pounds (1,816 kg) of that was normal bomb load, which it could carry to a service ceiling of but 17,000 feet (5,185 m). Fortunately for its crews, the Stirling proved quite manoeuvrable, and with eight machine guns, it had a respectable defensive armament. By mid 1943 the Stirling was being withdrawn from service. It could not fly as high as the new Lancasters and Halifaxes, and its small bomb bay could not accomodate the ever larger bombs being dropped by the RAF. The last Stirling bombing sortie took place in 1944,

The Stirling Mk.IV was stripped of armament and reconfigured as a glider tug and transport, doing quite well during the D-Day glider assaults. So successful was this effort that Shorts built all Mk.Vs as transports from the beginning. A faired-over tail turret position and a redesigned, streamlined nose, which served as an additional cargo hold, were included in modifications. Each Stirling V could carry 24 men with full equipment. The first Mk.Vs went into service with No. 46 Squadron, Transport Command, in February 1945.

Specification (Stirling III)

Powerplants: four Bristol Hercules XVI 1,650 hp 14-cylinder radials.

Dimensions: length 26.59 m (87 ft 3 in); height 6.93 m (22 ft 9 in); wing span 30.20 m (99 ft 1 in).

Weights: empty 21,293 kg (46,900 lb); operational 31, 751 kg (70,000 lb).

Performance: maximum speed 435 km/h (270 mph); service ceiling 5,180 m (17,000 ft); range 950 km (590 miles).

Armament: two .303 calibre machine guns each in nose and top turrets, four .303 guns in tail turret; internal bomb load 6,356 kg (14,000 lb).

Supermarine Spitfire Mk I-XIII

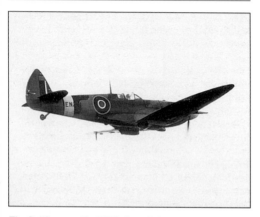

The Spitfire was the RAF's best fighter aircraft in 1940

Reginald Mitchell designed this most famous of British fighters from his experience with the outstanding S.6B Schneider Cup racing seaplanes, the fastest aircraft of their day. Flown for the first time on 5 March 1936, the Spitfire had such development potential that a total of 20,351 (plus 2,556 Seafires) were built with the last coming off the line in 1947. Though more Hurricanes fought in the Battle of Britain, the Spitfire became the symbol of Britain's will to resist a determined enemy at almost impossible odds. Pilots loved the aircraft with a passion which lasts to this day, saying it was more like pulling on a pair of pants than stepping into a machine.

The Merlin engine variants were considered by purists to be the definitive Spitfires with elegant lines, smaller dimensions and lighter handling. The definitive example, the Mk.IX, began as a stopgap with a hastily installed Merlin 61 to counter the Fw 190, which significantly outclassed the Spitfire Mk.V. As it turned out, just over 5,500 IXs were built and pilots, on the whole, thought it was the best of the breed. The Merlin aircraft were developed up to the Mk.XIII, a low-level photo recon version, and many were deployed as carrier-based Seafires. Several marks had their wings clipped for better roll rate but many thought this destroyed the look of the aircraft's elegant elliptical wings

Specification (Spitfire IX)

Powerplant: Rolls-Royce Merlin 61 1,560 hp 12-cylinder inline.
Dimensions: length 9.57 m (31 ft 4.5 in); height 3.63 m (11 ft 5 in); wing span 11.23 m (36 ft 10 in).
Weights: empty 2,547 kg (5,610 lb); operational 3,405 kg (7,500 lb).
Performance: maximum speed 656 km/h (408 mph); service ceiling 13,420 m (44,000 ft); range 1,060 km (660 miles).
Armament: two 20 mm cannon, four .303 calibre machine guns; external bomb load 454 kg (1,000 lb) or rockets.

Supermarine Spitfire Mk XIV-F.24

The Spitfire FRXVIII was further strengthed to accomodate extra fuel tanks in the wings and fuselage

The final versions of the Spitfire, from the Mk.XIV to the bubble canopy F.24 were powered by Rolls-Royce Griffon engines, some pumping out almost 2,500 hp and yielding top speeds of 460 mph. The aircraft's growth was phenomenal with a 100 per cent increase in power from the first mark to the last.

The Griffon Spits made ideal interceptors, particularly against V-1 flying bombs, with the first version, the Mk.XII, developed from the Mk.V airframe and going into service in the spring of 1943 on home defence. The Mk.XIV, based on the VIII airframe and entering service in January 1944, was still a stopgap version, though few complained as it had outstanding performance and handling. It also outmatched the Focke-Wulf FW 190 in all aspects except for the FW 190's unrivalled rate of roll. It was a Mark XIV that became the first Allied fighter to shoot down a Messerschmitt Me 262.

The first version designed specifically for the Griffon was the Mk.XVIII, which entered squadron service just as the war was ending. With the F.21, which flew combat from January 1945, the famous pure elliptical wing disappeared in favour of a modified shape with strengthened spars. The Mk.22 had a 12.34 m wingspan and six 20 mm cannon.

With the F.24 came an enlarged tail, inherited from the Spiteful. The lean, powerful looks of the Griffon Spitfires were quite some distance from their Merlin brothers but, according to many, just as beautiful. Some examples had contra-rotating propellers and the last examples were withdrawn from RAF service in 1954, closing a momentous period in aviation history.

Specification (Spitfire XIV)

Powerplant: Rolls-Royce Griffon 65 2,050 hp 12-cylinder inline.
Dimensions: 9.96 m (length 32 ft 8 in); height 3.89 m (12 ft 9 in); wing span 11.23 m (36 ft 10 in).
Weights: empty 3,042 kg (6,700 lb); operational 4,667 kg (10,280 lb).
Performance: maximum speed 721 km/h (448 mph); service ceiling 13,573 m (44,500 ft); range 740 km (460 miles).
Armament: two 20 mm cannon, four .303 calibre machine guns; external bomb load 454 kg (1,000 lb).

Vickers Wellington

A Wellington II of 12 Squadron, lost during the raid on Cologne on 11 October 1941

One of Britain's bomber hopefuls when it first flew in June 1936, the Wellington was one of the few which ended up fulfilling its initial promise. Construction was based on Barnes Wallis' geodetic design, which was initially difficult to apply to an operational aircraft but which, in the end, yielded immense strength and resistance to battle damage. When the war began it was the RAF's finest bomber with striking power and range to carry the war into Germany itself, the major selling point of Bomber

Command's existence through the early war years.

When the newer, four engine types began to replace it, the 'Wimpey,' fitted with sea search radar, was built primarily for Coastal Command's maritime mission. The aircraft's punch, with a search light, torpedoes, rockets and depth charges, was quite impressive. The final examples were built as transports and many of the older marks were used as bomber crew transition trainers. Honest, straightforward, even if somewhat stiff on the controls, the Wellington soldiered on after the war well into the 1950s, continuing to train crews in the art of navigation and bomber tactics. It will always be remembered as Bomber Command's mainstay in 1942, during the darkest days of the airwar against Germany.

Specification (Wellington III)

Powerplants: twin Bristol Hercules III 1,375 hp 14-cylinder radials.

Dimensions: length 21.30 m (64 ft 7 in); height 5.34 m (17 ft 6 in); wing span 25.06 m (82 ft 2 in).

Weights: empty 9,080 kg (20,000 lb); operational 13,393 kg (29,500 lb).

Performance: maximum speed 410 km/h (255 mph); service ceiling 5,790 m (19,000 ft); range 2,478 km (1,540 miles).

Armament: two .303 calibre machine guns in nose turret, four .303 guns in tail turret, two flexible .303 guns in beam positions; internal bomb load 2,043 kg (4,500 lb).

Bell P-39 Airacobra

One of the original Airacobras ordered by the RAF

Originally designed to the same specification as the P-38, the P-39 was supposed to have been a lightweight interceptor with performance in the 400 mph (644 km/h) range. Unique in having the engine mounted over the centre wing web behind the pilot, a nosewheel and automobile-style entry doors, the Airacobra's 37 mm cannon and six machine guns boded well. Unfortunately, increasing weight and a low altitude engine doomed the aircraft to mediocre performance. Flying against both the Germans in the

Mediterranean and the Japanese in New Guinea, P-39 squadrons were hacked apart.

The real success of the fighter came at the hands of the Russians who received over 5,000 and immediately put them to good use as ground attack aircraft. Universally, Soviet pilots loved their '39s as rugged, reliable aircraft able to pack a tremendous punch and several did quite well in air-to-air combat since much of the Eastern Front air war was fought at low level.

The ultimate development of the design came with the laminar flow wing P-63 Kingcobra, the fighter the P-39 should have been. It came too late for American squadrons but again the Russians got over 3,000 of them and did exceptionally well in combat with no complaints.

Specification (P-39Q)

Powerplant: Allison V-1710-85 1,325 hp 12-cylinder inline.
Dimensions: length 9.19 m (30 ft 2 in); height 3.61 m (11 ft 10 in); wing span 10.37 m (34 ft).
Weights: empty 2,563 kg (5,645 lb); operational 3,450 kg (7,600 lb).
Performance: maximum speed 605 km/h (376 mph); service ceiling 10,675 m (35,000 ft); range 1,730 km (1,075 miles).
Armament: four .50 calibre machine guns, one 37 mm cannon; one bomb up to 227 kg (500 lb).

Boeing B-17 Flying Fortress

A Boeing B-17E in pre-war scheme: later versions added a chin turret and extra machine guns

America's most famous aircraft, the B-17, was almost still-born. During the summer 1935 fly-off competition with the Douglas B-18, the control locks were not removed and the prototype Model 299 crashed, leaving victory to converted Douglas airliner. Fortunately, the promise of this heavy bomber brought an initial purchase of 13 test aircraft as coastal defence weapons, an extension of coastal artillery. An offensive weapon would never have got past an isolationist Congress. Though the RAF had very poor results with their early Fortresses, the USAAF built its daylight strategic bombing doctrine around the B-17, particularly in Europe.

Major development changes resulted in increasing numbers of defensive .50 calibre machine guns being mounted until the final version, the B-17G, carried 13 guns. In spite of such a formidable defence, machine guns were not enough to keep the Luftwaffe from almost halting the Eighth Air Force campaign from Britain in the autumn of 1943. The introduction of long-range escort fighters, particularly the P-51, finally allowed the daylight bombing of Hitler's Third Reich to continue until Germany was reduced to rubble. The Flying Fortress has, ever since the war, been the symbol of American resolve to win in the face of overwhelming odds.

Specification (B-17G)

Powerplants: four Wright R-1820-97 Cyclone 1,200 hp 9-cylinder radials.
Dimensions: length 22.8 m (74 ft 9 in); height 5.82 m (19 ft 1 in); wing span 31.62 m (103 ft 9 in).
Weights: empty 16,253 kg (35,800 lb); operational 29,782 kg (65,600 lb).
Performance: maximum speed 462 km/h (287 mph); service ceiling 10,675 m (35,000 ft); range 1,770 km (1,100 miles).
Armament: two .50 calibre machine guns each in chin, top, ball and tail turrets; five flexible .50 calibre machine guns; internal bomb load 2,724 kg (6,000 lb) with optional external bomb racks for increased load.

Boeing B-29 Superfortress

Huge formations of Boeing B-29s effectively destroyed Japanese industry in early 1945

For U.S. Army Air Forces' chief Gen. Hap Arnold, the B-29 was the central proof and hope of air power's ability to hit the enemy at its heart. As a result, he pushed the Superfortress into service long before it was ready, leading to numerous problems, the most serious being engine fires. Ordered in August 1940, the first prototype flew in September 1942, nine months after a production contract was signed for 500 aircraft. Not only was Boeing to build as many of the new pressurized, high altitude bombers as possible, but Bell, North American and

General Motors were contracted to open production lines.

The first combat missions flown out of China against Japan in June 1944 were not very successful, so the effort was moved to the newly captured Marianas Islands late in the year with the first mission against Tokyo being flown on November 24th. By early 1945 over 500 Superforts would head for Japan on a single mission and Japan began to burn to the ground, making the B-29 the single weapon most feared by the enemy. The ultimate example of the aircraft's striking power was the dropping of the atomic bombs on Hiroshima and Nagasaki in August 1945, bringing peace and avoiding the need for a ground invasion of Japan.

Specification (B-29)

Powerplants: four Wright R-3350-23 Duplex Cyclone 2,200 hp 18-cylinder radials.
Dimensions: length 30.18 m (99 ft); height 8.46 m (27 ft 9 in); wing span 43.05 m (141 ft 3 in).
Weights: empty 33,823 kg (74,500 lb); operational 61,290 kg (135,000 lb).
Performance: maximum speed 574 km/h (357 mph); service ceiling 10,980 m (36,000 ft); range 5,229 km (3,250 miles).
Armament: two .50 calibre machine guns each in four remotely controlled turrets, two .50 calibre guns and one 20 mm cannon in tail turret; internal bomb load 9,080 kg (20,000 lb).

Brewster F2A Buffalo

A Brewster Buffalo in the orange markings of the Dutch air force, seen over the Dutch East Indies

Though the F2A won the competition against the F4F Wildcat for the U.S. Navy's first monoplane fighter, the Buffalo was a big disappointment and the Wildcat was subsequently ordered as insurance, a fortunate decision. Several other nations, including Finland, Belgium, Britain, and Holland, ordered the aircraft to bolster their struggling air arms. Of all the users, only the Finns seemed to find their Buffaloes effective, flying them in combat with excellent results from February 1940 through to the end of the war, producing several aces. Some ex-Belgian Buffaloes were among the handful of RAF fighters available during the last stages of the campaign in Greece.

When World War II began in the Pacific, most Navy fighter squadrons were flying the Wildcat but VF-2, VF-3 and VFM-221 flew Buffaloes against the Japanese as did five RAF and one Australian squadron. The tubby fighter was quickly blasted from the arena by the superior A6M Zero-Sen. The British attempted to lighten their Buffaloes by removing ammunition and fuel, and installing lighter guns, in order to increase its performance, but it made little difference. The unfortunate Buffalo was overweight, underpowered and simply too unmanoeuvrable to survive in combat against the very agile Japanese fighters and their highly trained pilots.

Specification (F2A-2)

Powerplant: Wright R-1820-40 Cyclone 1,100 hp 9-cylinder radial.

Dimensions: length 8.03 m (26 ft 4 in); height 3.68 m (12 ft 1 in); wing span 10.67 m (35 ft).

Weights: empty 2,102 kg (4,630 lb); operational 3,178 kg (7,000 lb).

Performance: maximum speed 483 km/h (300 mph); service ceiling 9,303 m (30,500 ft); range 1,553 km (965 miles).

Armament: four .30 or .50 calibre machine guns.

Consolidated B-24 Liberator

A B-24D of the first production run: over 18,000 B-24s were manufactured during World War II

First flown in December 1939, the B-24 ended World War II as the most produced American aircraft in history at over 18,000 examples, thanks in large measure to Henry Ford and the harnessing of American industry. Overshadowed in publicity by its stablemate, the B-17, the Liberator served in every theatre of war with almost every Allied nation. Just a little faster with a slightly larger bomb load and longer range than the Fort, the Lib had a lower service ceiling and ended up getting hit by fighters and flak more often. Nevertheless, the loss rates of the two aircraft were too close to say one was more vulnerable than the other.

One thing was certain, the B-24 was much heavier on the controls and harder to hold in tight formation, creating an enormous workload for its pilots. Its high fuselage-mounted Davis wing also meant it was dangerous to ditch or belly land since the fuselage tended to break apart. The Liberator's most famous mission was the low-level strike against the Ploesti oil fields in Rumania, 1 August 1943, which unfortunately turned into a disaster due to attack waves getting out of sequence.

The US Navy flew the PB4Y patrol version, ranging far across the Atlantic from Iceland as well as the USA. Many were built as C-87 transports, F-7 photo reconnaissance platforms and C-109 fuel tankers. The B-24 was one of the most important Allied aircraft of World War II.

Specification (B-24J)

Powerplants: four Pratt & Whitney R-1830-65 Twin Wasp 1,200 hp 14-cylinder radials.
Dimensions: length 20.47 m (67 ft 2 in); height 5.49 m (18 ft); wing span 33.53 m (110 ft).
Weights: empty 16,798 kg (37,000 lb); operational 29,510 kg (65,000 lb).
Performance: maximum speed 467 km/h (290 mph); service ceiling 8,540 m (28,000 ft); range 3,540 km (2,200 miles).
Armament: two .50 calibre machine guns each in nose, top, ball and tail turrets; two .50 calibre guns in waist positions; internal bomb load 3,632 kg (8,000 lb) with optional external bomb racks.

Consolidated PBY Catalina

A Catalina of the Royal New Zealand Air Force

In spite of being ordered in late 1933 and first flying in March 1935, the Catalina, as both a flying boat and an amphibian, became one of the durable and more effective aircraft of World War II. The "Cat" often had more range than crews wanted.

Doing just over 100 mph (161 km/h) in cruise, it could fly close to 3,000 miles (4,827 km), something between 25 to 30 hours in the air, well beyond the endurance of even those aircraft with two sets of pilots. Yet, it was not unusual to fly well into exhaustion under the pressures of war when sighting the enemy could mean the difference between victory or defeat. It was a PBY that reported the first sighting of the Japanese fleet approaching Midway. Such scouting missions could be incredibly dangerous as Japanese carrierborne aircraft could make short work of a PBY if they caught it.

Quickly nicknamed "Dumbo" for their resemblance to Walt Disney's flying, floppy-eared cartoon elephant, rescue Catalinas (including Army OA-10As) were the most loved of all aircraft in any theatre. Before long any rescue aircraft became a Dumbo regardless of type. Several squadrons were equipped with torpedoes, painted their Cats black and flew effective night shipping strikes. By the time the war ended more Catalinas had been built than any single seaplane in history, even though faster and better equipped aircraft were available.

Specification (PBY-5A)

Powerplants: twin Pratt & Whitney R-1830-92 Twin Wasp 1,200 hp 14-cylinder radials.

Dimensions: length 19.47 m (63 ft 10 in); height 6.5 m (20 ft 2 in); wing span 31.7 m (104 ft).

Weights: empty 9,485 kg (20,910 lb); operational 15,416 kg (33,975 lb).

Performance: maximum speed 290 km/h (180 mph); service ceiling 4,480 m (14,700 ft); range 4,096 km (2,545 miles).

Armament: three flexible .50 calibre machine guns; external racks for bombs, torpedoes or depth charges.

Curtiss P-40 Tomahawk/Kittyhawk/Warhawk

A Curtiss P-40N Warhawk, the final production version

Although the Curtiss P-36 Hawk was already out of date, the company chose the quick way to enter the Army fighter competition of May 1939 with the XP-40, nothing more than an inline Allison engined version of the same aircraft. In spite of its built-in obsolescence, the P-40 was ordered in massive numbers, the largest American fighter buy in history, just in time to have enough on hand at the beginning of World War II. Famously painted with sharkmouths, the RAF in the Western Desert, Claire Chennault's Flying Tigers and Army pilots in the Pacific held the superior Bf 109, Fw 190, Zero and

Oscar at bay: an accomplishment which makes the aircraft one of the most important fighters of World War II.

A succession of models led to a variety of names, often confusing even those who flew it. The early models through to the P-40C were named Tomahawk, the next series through to the E were named Kittyhawk and the final versions were called Warhawk. Many, if not most, U.S. Army fighter pilots in China considered the P-40 more effective than the P-51 for ground support and air combat in the theatre.

In spite of being taken off the production line in 1944, the P-40 remained in combat through to the end, though most were assigned to training units in the US for fighter transition.

Specification (P-40E)

Powerplant: Allison V-1710-39 1,150 hp 12-cylinder inline.
Dimensions: length 9.51 m (31 ft 2 in); height 3.23 m (10 ft 7 in); wing span 11.39 m (37 ft 4 in).
Weights: empty 2,883 kg (6,350 lb); operational 3,759 kg (8,280 lb).
Performance: maximum speed 583 km/h (362 mph); service ceiling 8,845 m (29,000 ft); range 1,368 km (850 miles).
Armament: six .50 calibre machine guns; one 227 kg (500 lb) and two 45 kg (100 lb) bombs.

Curtiss SB2C Helldiver

The SB2C-4 Helldiver failed to live up to expectations

The company's first production monoplane bomber in a long line of previously successful aircraft named Helldiver, the prototype SB2C crashed just a few weeks after its first flight . Full-scale production had been ordered on 29 November 1940, before anyone knew how it flew, which was unfortunate because it turned out to fly more like a contemporary garbage truck. From the results gained during the short time the prototype flew, drastic changes were ordered, including an enlarged fin and rudder, increased fuel capacity, self-sealing fuel tanks, fixed armament doubled to four .50-calibre wing guns and more armour.

The first production Helldiver did not roll out until June 1942, with the first U.S. Navy squadron getting their aircraft in December. Another 11 long months was spent trying to get the SB2C operational. It earned its nickname, "The Beast." Neither pilots nor aircraft carrier skippers seemed to like it. A few were supplied to the Royal Navy Fleet Air Arm, but they did not see action,

The Truman Committee investigated Helldiver production and turned in a scathing report, probably the beginning of the end for Curtiss. In spite of its problems, the Helldiver was flown through the last two years of the Pacific War with a fine combat record due to the high training of its crews.

Specification (SB2C-3)

Powerplant: Wright R-2600-20 Double Cyclone 1,900 hp 14-cylinder radial.

Dimensions: length 11.21 m (36 ft 9 in); height 4.5 m (14 ft 9 in); wing span 15.16 m (49 ft 9 in).

Weights: empty 4,592 kg (10,114 lb); operational 6,208 kg (13,674 lb).

Performance: maximum speed 475 km/h (294 mph); service ceiling 7,625 m (25,000 ft); range 1,930 km (1,200 miles).

Armament: two forward firing 20 mm cannon, two rear firing flexible .50 calibre machine guns; internal bomb load 454 kg (1,000 lb) with two external racks for additional load.

Douglas A-20 Havoc/Boston

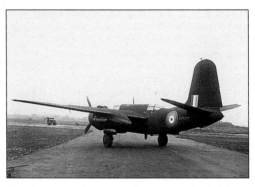

The Havoc served as a very effective night fighter

The Douglas A-20 Havoc was as close as a bomber pilot could get to being a fighter pilot in 1942. A single-seat cockpit about the size of a fighter's, decent speed at 350 mph (563 km/h) (at least in the early models before they started adding weight), and light controls for a bomber all added up to excellent performance. The design was produced initially in 1938 as the DB-7 Boston for foreign buyers, the French and British in particular, before the U.S. Army ordered it as the A-20 Havoc. By the time the war ended, the aircraft was the most produced in the attack classification.

The A-20C was the first try at standardizing American and British requirements into one version. Fitted with 1,600 hp R-2600s, the aircraft's gross weight went from 20,711 pounds (9,403 kg) to 25,600 pounds (11,622 kg), bringing the top speed down to 342 mph (550 km/h). On the deck, an Fw 190A was hardly faster. This was the first version to see action, flying with the 15th Bomb Squadron, which initiated combat operations for the Eighth Air Force on 4 July 1942. Several Havocs were turned into night fighters and the RAF kept using them for low-level strike. In the Pacific low-flying A-20s did enormous damage to Japanese shipping and ground installations.

Specification (A-20G)

Powerplants: twin Wright R-2600-23 Double Cyclone 1,600 hp 14-cylinder radials.

Dimensions: length 14.63 m (48 ft); height 5.36 m (17 ft 7 in); wing span 18.69 m (61 ft 4 in).

Weights: empty 7,809 kg (17,200 lb); operational 12,258 kg (27,000 lb).

Performance: maximum speed 510 km/h (317 mph); service ceiling 7,625 m (25,000 ft); range 1,650 km (1,025 miles).

Armament: six forward firing .50 calibre machine guns, two .50 calibre guns in top turret, one flexible .50 calibre gun in ventral position; internal bomb load 908 kg (2,000 lb), external bomb load 908 kg (2,000 lb).

Douglas A-26 Invader

The A-26 invader was so successful it survived to fight again in Korea and Vietnam

The Douglas A-26 Invader was designed to a 1940 Army Air Corps specification and it turned out to be the last operational aircraft in the attack series of fast, light tactical bombers. From the first flight of the XA-26 on 10 July 1942, the Invader was, like the company's earlier A-20, a pilot's airplane, more like a fighter than a bomber, with a single-pilot cockpit. In spite of its size and power, it was easy to fly and everything was within easy reach as one would expect from a fighter. The first A-26Bs went into combat with the Ninth Air Force over Europe in November 1944. By the end of the war several units in the Pacific had made the transition to Invaders.

The top speed of the A-26C was over 370 mph (595 km/h) and high cruise was 284 mph (457 km/h), permitting operations without escort. On those missions with escort, fighters did not have to weave or pull the power back, something they much appreciated. With 4,000 pounds (1,816 kg) of bombs and as many as 14 forward firing .50 calibre machine guns in the nose and in external packages, the A-26 was a potent weapon. Many other weapon fits were tested, including the substitution of a 75 mm gun or two 37 mm cannon for the six machine guns in the nose of the A-36B.

There was enough capability in the Invader to carry it through two more wars, Korea and Vietnam, before it was finally retired in the late 1960s, speaking reams about its fine design.

Specification (A-26B)

Powerplants: twin Pratt & Whitney R-2800-71 Double Wasp 2,000 hp 18-cylinder radials.
Dimensions: length 15.24 m (50 ft); height 5.64 m (18 ft 6 in); wing span 21.34 m (70 ft).
Weights: empty 10,156 kg (22,370 lb); operational 14,528 kg (32,000 lb).
Performance: maximum speed 571 km/h (355 mph); service ceiling 6,735 m (22,100 ft); range 2,253 km (1,400 miles).
Armament: six to eight forward firing .50 calibre machine guns, two .50 calibre guns each in remotely controlled top and bottom turrets; internal bomb load 1,816 kg (4,000 lb), optional external load 908 kg (2,000 lb).

Douglas C-47 Dakota/Skytrain

The C-47 was one of the most important Allied aircraft of World War II

The Douglas C-47 was considered one of the most important weapons of World War II by such commanders as Eisenhower and MacArthur. Taken straight off the civil transport production line, the DC-3 was inducted into the military by inserting a large cargo door and camouflaging it; no armour or self-sealing fuel tanks were added, making it one of the most vulnerable of all wartime aircraft.

Nevertheless, few aircraft were better suited than the "Gooney Bird" for landing and taking off at the short forward airstrips at the front lines; just about anything one could cram inside, it could carry, unless an engine failed.

Later in the war, it became crucial for successful invasion of enemy territory, hauling paratroops and towing gliders, then resupplying forward areas near the front line.

Pilots found it very forgiving in even the roughest of weather and field conditions, easy to handle and manage. In spite of having less capacity and less power than its younger brothers (like the C-46 and C-54), the C-47 was used more often and in more combat action than any single Allied transport.

Specification (C-47A)

Powerplants: twin Pratt & Whitney R-1830-92 Twin Wasp 1,050 hp 14-cylinder radials.
Dimensions: length 19.44 m (63 ft 9 in); height 5.19 m (17 ft); wing span 29.13 m (95 ft 6 in).
Weights: empty 8,111 kg (17,865 lb); operational 14,074 kg (31,000 lb).
Performance: maximum speed 370 km/h (230 mph); service ceiling 8,052 m (26,400 ft); range 5,792 km (3,600 miles).
Armament: none.

Douglas SBD Dauntless

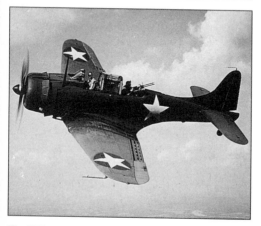

The SBD was also ordered by the US Army as the A24; note the absence of an arrestor hook

Considered obsolete even before the attack on Pearl Harbor, the SBD Dauntless was supposed to give way to the bigger, faster Curtiss SB2C Helldiver, which turned out to be a disappointment. As things turned out, the "Slow But Deadly" SBD and the Wildcat ended up being the only pre-Pearl Harbor U.S. carrier aircraft still in service by the end of the war. In those grim days from May to November 1942, Navy

Dauntlesses and Wildcats pulled off the miracle of destroying most of the Japanese carrier fleet, the high point being the Battle of Midway.

Marine SBDs played a crucial role in holding Guadalcanal. Once, a single Dauntless was the only attack aircraft available to the island's defenders. This humble-looking aircraft took part in all five naval engagements fought only by aircraft carriers. Pilots always wanted more speed and more armament but that didn't stop SBDs from turning the tide of the Pacific War. In spite of its primary role as a dive bomber, the small aircraft was so manoeuvrable, many Dauntless pilots and their rear gunners shot down quite a number of Japanese fighters with both forward firing and rear firing guns.

Specification (SBD-3)

Powerplant: Wright R-1820-52 Cyclone 1,000 hp 9-cylinder radial.

Dimensions: length 10.07 m (33 ft .125 in); height 3.94 m (12 ft 11 in); wing span 12.66 m (41 ft 6.25 in).

Weights: empty 3,030 kg (6,675 lb); operational 4,928 kg (10,855 lb).

Performance: maximum speed 394 km/h (245 mph); service ceiling 7,412 m (24,300 ft); range 1,770 km (1,100 miles).

Armament: two forward firing .50 calibre machine guns, two rear firing flexible .30 calibre guns; 1,200 lb (545 kg) bomb load.

Grumman F4F/FM Wildcat

One of the F4Fs ordered by France but delivered to the RAF instead after the French surrender

Though the Brewster F2A Buffalo beat out the Grumman F4F Wildcat in the initial competition for the Navy's first monoplane fighter, the loser was developed and eventually proved to be far superior. The type was ordered into mass production, thankfully before the attack on Pearl Harbor. The Wildcat's landing gear was retractable but the pilot had to change hands on the stick and crank 29 turns on a small handcrank which was connected to a bicycle chain.

"A beer barrel on a roller skate run through with an ironing board," as it was described, the F4F was outclassed in speed, manoeuvrability and range by the Japanese A6M2 Zero-Sen, but it could absorb battle damage, had superior armour protection, could dive away from the Zero and had excellent .50 calibre machine guns to chop up light enemy fighters and bombers with little problem. Through superior tacticians like Jimmy Thach, who invented the mutual covering "Thach Weave," the F4F became a most formidable fighter, holding the fort in the darkest days of the Pacific War along with the Army's P-40. So successful was the design for escort carriers, production was turned over to General Motors' Eastern Aircraft as the FM-2, which was built until the surrender.

Specification (F4F-4)

Powerplant: Pratt & Whitney R-1830-86 Twin Wasp 1,200 hp 14-cylinder radial.
Dimensions: length 8.76 m (28 ft 9 in); height 3.63 m (11 ft 11 in); wing span 11.58 m (38 ft).
Weights: empty 2,111 kg (4,649 lb); operational 2,769 kg (6,100 lb).
Performance: maximum speed 512 km/h (318 mph); service ceiling 10,675 m (35,000 ft); range 1,448 km (900 miles).
Armament: six .50 calibre machine guns; wing racks for two 113 kg (250 lb) bombs.

Grumman F6F Hellcat

Nearly three-quarters of the US Navy's air-to-air kills in the Pacific were achieved by F6Fs

Though not as fast, at 380 mph (611 km/h), as other American late war fighters, the F6F carried excellent firepower and had that legendary Grumman Iron Works rugged-as-a-bridge construction. Developed as a follow-on to the F4F, the Hellcat was a significant step forward in carrier aviation, ideal for long periods at sea. A testimony to its basic design, very few changes were made to the basic airframe throughout its substantial production run of over 12,000 machines. By the end of the war the Hellcat, along with the Corsair, dominated the Pacific carrier war.

Pilots had so much confidence in the aircraft and themselves they complained of not having enough enemy aircraft to hunt down.

From the time the Hellcat entered combat in August 1943, in almost exactly two years it shot down 5,174 enemy aircraft, almost 75% of all the Navy's air-to-air kills. Simple to maintain, easy to fly, particularly when coming back aboard ship, the F6F was safe, rugged, able to take a significant amount of battle damage and still come back to the carrier or land bases. Two different night fighter versions were developed and it was also flown by the Royal Navy and from land bases by both U.S. Navy and Marine pilots. The F6F was a perfect example of a simple design being in the right place at exactly the right time.

Specification (F6F-3)

Powerplant: Pratt & Whitney R-2800-10W Double Wasp 2,000 hp 18-cylinder radial.

Dimensions: length 10.24 m (33 ft 7 in); height 4 m (13 ft 1 in); wing span 13.06 m (42 ft 10 in).

Weights: empty 4,105 kg (9,042 lb); operational 5,532 kg (12,186 lb).

Performance: maximum speed 605 km/h (376 mph); service ceiling 11,438 m (37,500 ft); range 1,754 km (1,090 miles).

Armament: six .50 calibre machine guns; underwing racks for six rockets or 908 kg (2,000 lb) bomb load.

Grumman TBF/TBM Avenger

The TBF first saw action at the battle of Midway where US torpedo bombers suffered heavy losses

A replacement for the obsolete TBD Devastator, the first TBF Avenger was delivered to the U.S. Navy in January 1942. The torpedo bomber went into action with Torpedo Eight (VT-8) during the Battle of Midway the following June, though the new type was overshadowed by the squadron's loss of all its TBDs and most of its TBFs during the battle. The large carrier-based machine quickly established itself as an excellent aircraft, though the life of a torpedo bomber crew was very short.

The Avenger served through World War II, toward the end primarily in the guise of General Motors' TBM counterpart. GM's Eastern Aircraft Division, taking on production of the Avenger and the Wildcat, freed Grumman to build record numbers of F6F Hellcats. Many considered the torpedo bomber mission near suicide, something born out with devastating reality during long, slow, low-level runs at enemy ships. Nevertheless, the Avenger soldiered on with a creditable survival rate, particularly when its primary mission was changed to bombing and attack against land targets. Though extremely stiff on the controls, its stability was an excellent trait during attack runs.

Specification (TBF-1)

Powerplant: Wright R-2600-8 Double Cyclone 1,700 hp 14-cylinder radial.

Dimensions: length 12.19 m (40 ft); height 5 m (16 ft 5 in); wing span 16.51 m (54 ft 2 in).

Weights: empty 4,576 kg (10,080 lb)s; operational 6,205 kg (13,667 lb).

Performance: maximum speed 436 km/h (271 mph); service ceiling 6,527 m (21,400 ft); range 1,955 km (1,215 miles).

Armament: one forward firing .30 calibre machine gun, one .50 calibre gun in rear turret; one flexible .30 calibre gun in ventral position; internal bomb load 908 kg (2,000 lb) of bombs, torpedo or depth charges.

Lockheed Hudson

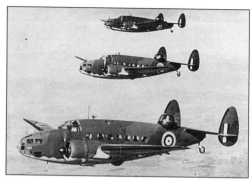

The RAF ordered the Hudson on the eve of World War II

When Lockheed designer Kelly Johnson said he could produce a recon-bomber from the Model 14 airliner, company executives asked the British Purchasing Commission to inspect a mockup. The British were impressed enough to order 200 Model B14Ls right off the drawing board on 23 June 1938. A Hudson flew for the first time on 10 December 1938 and 1,500 went to the RAF and RAAF under the Commission contract before they were included in Lend-Lease procurement and received the Army Air Corps A-28 designation, and later A-29 with new engines.

In combat with the RAF from the beginning of World War II, claiming the first German aircraft kill, the Hudson quickly earned the nickname 'Old Boomerang' for its ability to come back with extensive battle damage. After being replaced by more modern bombers, the Hudson kept flying combat in a number of secondary roles, including secret agent delivery to and from the continent. All USAAF Hudsons were built with RAF camouflage and serial numbers, but the US army requisitioned quite a few after Pearl Harbor for bomber crew training and coastal anti-submarine patrol. A USAAF A-29 made the first successful American attack on a U-boat in World War II. The bomber was developed into the more powerful PV patrol bomber series.

Specification (Hudson I)

Powerplants: twin Wright GR-1820-G10A Cyclone 1,100 hp 9-cylinder radials.
Dimensions: length 13.51 m (44 ft 4 in); height 3.62 m (11 ft 10.5 in); wing span 19.96 m (65 ft 6 in).
Weights: empty 5,448 kg (12,000 lb); operational 8,399 kg (18,500 lb).
Performance: maximum speed 396 km/h (246 mph); service ceiling 7,473 m (24,500 ft); range 3,154 km (1,960 miles).
Armament: five flexible .303 calibre machine guns, two .303 guns in top turret; internal bomb load 340 kg (750 lb).

Lockheed P-38 Lightning

Both America's top-scoring fighter aces achieved their victories in the P-38

The most advanced aircraft of its day when it first flew on 27 January 1939, Kelly Johnson's P-38 became one of America's more famous aircraft in World War II as both a fighter and a photo recon platform. Early models of the P-38 had streamlined engine cowlings and smaller radiators, virtually no cockpit heat and turbosupercharger problems. Nevertheless, these first Lightnings made the aircraft's reputation as a multirole aircraft. Pilots found they could outclimb and outdistance the opposition, giving them the key ability to choose the time and place for combat. In late 1942, P-38s became the first modern American fighters to best the opposition.

With counter rotating propellers, concentrated firepower, twin-engine safety, hydraulically boosted ailerons and range, the Lightning was in a class of its own and it was the mount of America's two leading aces, Dick Bong (40 kills) and Tommy McGuire (38 kills). It was so streamlined it had problems with compressibility, going fast enough in a dive to lock up the controls.

Specification (P-38J)

Powerplants: twin Allison V-1710-89/91 1,425 hp 12-cylinder inlines.

Dimensions: length 11.53 m (37 ft 10 in); height 3.91 m (12 ft 10 in); wing span 15.85 m (52 ft).

Weights: empty 5,797 kg (12,780 lb); operational 7,945 kg (17,500 lb).

Performance: maximum speed 666 km/h (414 mph); service ceiling 13,420 m (44,000 ft); range 3,636 km (2,260 miles).

Armament: four .50 calibre machine guns, one 20 mm cannon; external bomb load 1816 kg (4,000 lb).

Lockheed PV-1 Ventura/PV-2 Harpoon

Lockheed PV-1 Venturas serving with the Royal New Zealand Air Force

Largely overlooked as a weapon of war, the Lockheed PV-1 Ventura was developed for the British from the Hudson bomber and Lodestar transport. With 2000 hp engines the Ventura was more of a hotrod than a bomber to its crews and could carry a naval torpedo internally. Overshadowed by more publicized Navy patrol bombers, the PV was one of the fastest at 312 mph (502 km/h), serving in both the RAF and the U.S. Navy.

Over the Pacific, several pilots chased and shot down Zeros with their forward firing machine guns.

However, like its older Lockheed brothers, the PV was a real handful on the ground with inadequate vertical surfaces and rudders. Nevertheless, pilots enjoyed it far more than the slower (282 mph (454 km/h) max) and more docile PV-2 Harpoon successor which had longer wings and larger horizontal and vertical tail surfaces. Venturas did some of their finest work over the vast stretches of the Pacific and on the Empire Express run in the Aleutians.

The PV-2 Harpoon entered combat in early 1944, carrying out naval patrol bomber missions with an increased payload and longer range. The Navy ordered 500 PV-2s of which 470 were delivered by the time of the Japanese surrender.

Specification (PV-1)

Powerplants: twin Pratt & Whitney R-2800-31 Double Wasp 2,000 hp 18-cylinder radials.

Dimensions: length 15.67 m (51 ft 5 in); height 4.02 m (13 ft 2 in); wing span 19.96 m (65 ft 6 in).

Weights: empty 8,795 kg (19,373 lb); operational 14,108 kg (31,077 lb).

Performance: maximum speed 502 km/h (312 mph); service ceiling 8,022 m (26,300 ft); range 2,670 km (1,660 miles).

Armament: five forward firing .50 calibre machine guns, two .50 calibre guns in top turret, two .50 calibre guns in ventral position; internal bomb load 1,135 kg (2,500 lb) bombs, depth charges or torpedo.

Martin B-26 Marauder

The first unescorted B-26 missions over France led to heavy casualties

The B-26 Marauder got an undeserved reputation as a 'Widow Maker' after entering U.S. Army service in February 1941. The first 'by the numbers' aircraft in American military service, the Marauder had to be flown by exact airspeeds, particularly on final approach and when one engine was out. These numbers – 150 mph (241 km/h) on short final – were intimidating to pilots who were used to far lower speeds, and whenever they slowed down below what the manual stated they usually stalled out and crashed. Once crews were re-trained, the Marauder became a safer aircraft.

Early in B-26B production, modifications were

made for a six-foot wing span increase, a larger fin and rudder and four .50-calibre machine gun fuselage side packs. The aerodynamic changes had come after the early spate of crashes. With the B-26F came a 3.5 degree increase in wing incidence angle to give the aircraft improved take-off performance. The Marauder went through a number of design changes throughout its service life, but this visible difference in the raised thrust line of its Pratt & Whitney R-2800 engines is one of the easiest to spot, a characteristic of all subsequent models. Clearing up its early bad reputation, the Marauder ended World War II with the lowest loss rate of any USAAF bomber.

Specification (B-26B)

Powerplants: twin Pratt & Whitney R-2800-39 Double Wasp 2,000 hp 18-cylinder radials.

Dimensions: length 17.77 m (58 ft 3 in); height 5.82 m (19 ft 1 in); wing span 19.83 m (65 ft) (through early B), 21.64 m (71 ft) (remaining).

Weights: empty 10,442 kg (23,000 lb); operational 17,327 kg (38,200 lb).

Performance: maximum speed 499 km/h (310 mph); service ceiling 17,327 m (23,000 ft); range 1,850 km (1,150 miles).

Armament: four forward firing .50 calibre machine guns, two .50 calibre guns each in top and tail turrets, four flexible .50 calibre guns; internal bomb load 2,630 kg (5,200 lb).

Martin Maryland/Baltimore

This Maryland obtained by the RAF served until 1942 when damaged belly-landing at Takali, Malta

Seldom recognized by the general public as American aircraft, Martin Maryland and Baltimore bombers served with distinction in numerous RAF squadrons. The Martin 167W with 1,200 hp engines was originally built for a U.S. Army Air Corps attack competition but after flying in February 1939 it was offered solely for export. The French Purchasing Commission ordered several but the order was taken over by the British after the defeat of May 1940 and the aircraft was named the Maryland. Ironically, several which had been delivered to France were flown by the Vichy forces against the Allies, who were also flying Marylands.

The type was successful enough in the Middle East for the British to ask for an advanced version, which became the Baltimore with 1,600 hp engines and a deepened fuselage to carry a larger bomb load. The first one flew on 14 June 1941. Used only in the Mediterranean Theatre, Baltimores entered combat with No.223 Squadron, RAF, in January 1942 and proved to be excellent light bombers, very similar to the A-20 in both top speed (just over 300 mph (483 km/h)) and bomb load (2,000 pounds (908 kg)). The aircraft was popular enough to be used by other nations, including Australia, Italy, South Africa and Turkey.

Specification (Baltimore III)

Powerplants: twin Wright R-2600-19 Double Cyclone 1,660 hp 14-cylinder radials.

Dimensions: length 14.79 m (48 ft 6 in);height 5.41 m (17 ft 9 in); wing span 18.69 m (61 ft 4 in).

Weights: empty 6,900 kg (15,200 lb); operational 10,442 kg (23,000 lb).

Performance: maximum speed 486 km/h (302 mph); service ceiling 7,320 m (24,000 ft); range 1,705 km (1,060 miles).

Armament: four forward firing .303 calibre machine guns, two to four .303 calibre guns in top turret, two .303 calibre guns in ventral position; internal bomb load 999 kg (2,200 lb).

North American B-25 Mitchell

One of the B-25s supplied to the USSR, but still carrying its original USAAF serial number

The B-25 Mitchell was about as close to an ideal medium bomber as any planner or pilot could hope for, adaptable in the extreme. Certainly Doolittle's Tokyo Raiders proved it could do the near impossible by flying off the USS *Hornet*, but innovators like Col. Pappy Gunn in the Pacific proved it could not only bomb but strafe when field modified with forward-firing .50 calibres in place of a bombardier. This devastating modification was standardized on the production line to incorporate two, four or eight .50s in a solid nose and (often) four more in fuselage sidepacks.

Both the B-25G and H were fitted with a single 75

mm cannon, another Pappy Gunn innovation, but this misplaced artillery piece was never very effective, particularly since it required a crewman to manually reload it each time a round was fired.

Of all American twin-engine bombers, the Mitchell was the easiest to fly and land. Stable, yet manoeuvrable when enough muscle was applied, the aircraft could be pointed and held in place, whether delivering bombs or firing guns. Such ease of operation allowed the pilot to give his attention earlier to the more important aspects of combat flying. Mitchells were provided to the RAF and the Soviet air arm and were widely exported to South and Central America after 1945.

Specification (B-25J)

Powerplants: twin Wright R-2600-29 Double Cyclone 1,850 hp 14-cylinder radials.

Dimensions: length 16.13 m (52 ft 11 in); height 4.80 m (15 ft 9 in); wing span 20.60 m (67 ft 7 in).

Weights: empty 9,579 kg (21,100 lb); operational 15,876 kg (35,000 lb).

Performance: maximum speed 443 km/h (275 mph); service ceiling 7,320 m (24,000 ft); range 2,413 km (1,500 miles).

Armament: (glass nose) five or six forward firing .50 calibre machine guns, two .50 calibre guns each in top and tail turrets, three flexible .50 calibre guns; internal bomb load 1,816 kg (4,000 lb).

North American P-51 Mustang

The P-51 had the range to escort US bombers to Germany and back from bases in the UK

When the British were hard-pressed during the first year of World War II they came to the USA asking manufacturers to build aircraft for them. Though the small North American Company was requested to build P-40s under licence they countered with a proposal for an entirely new fighter, the NA-73X, which became the P-51 Mustang.

The first Mustang Mk.Is entered combat in mid 1942 as ground attack and recce aircraft. The aircraft was also modified for the US Army to become the A-36 Invader dive bomber. Though very fast at low level with excellent handling, the early Allison-powered Mustangs lacked high altitude performance so the fighter was modified for the Rolls-Royce Merlin which turned it into the P-51B high altitude escort fighter. Prototypes exceeded the climb rate of the P-38 Lightning and the USAAF immediately ordered large numbers. With extremely long range and the ability to fight the enemy on equal terms in his own back yard, the Merlin Mustangs entered combat in December 1943, escorting US bombers attacking the German naval base at Kiel. Featuring a near perfect cockpit and excellent bubble canopy, the subsequent P-51D reigned supreme across the globe from mid 1944 until the end of the war. Total production exceeded 14,000.

In 1950 the Mustang, redesignated F-51D, was pulled out of mothballs for ground support in the Korean War where, unfortunately, its liquid cooling system was too vulnerable to ground fire.

Specification (P-51D)

Powerplant: Rolls-Royce Packard-built V-1650-7 Merlin 1,590 hp 12-cylinder inline.

Dimensions: length 9.82 m (32 ft 2.5 in); height 4.17 m (13 ft 8 in); wing span 11.3 m (37 ft .5 in).

Weights: empty 3,232 kg (7,125 lb); operational 5,266 kg (11,600 lb).

Performance: maximum speed 703 km/h (437 mph); service ceiling 12,780 m (41,900 ft); range 2,092 km (1,300 miles).

Armament: six .50 calibre machine guns, external bomb load 908 kg (2,000 lb) or drop tanks.

Northrop P-61 Black Widow

Designed from the outset as a night fighter, the P-61 was one of the biggest fighters of the war

With data from early RAF night fighter operations, Northrop's P-61 Black Widow was the first aircraft designed from the outset as a night fighter. Other night fighters, including the P-70 derivative of the A-20, the P-38M and the Bf 110, were modified versions of existing aircraft. When the XP-61 flew in May 1942 it was most impressive, as large as many medium bombers. The result was potent indeed, with four 20 mm cannon, four 12.7 mm (.50 in) machine guns, a crew of three (pilot, radar observer and gunner) and two 2,000 horsepower engines. The first P-61s went into service in May 1944, getting their first kills in the Pacific during July. In the same month, P-61s shot down four German bombers in their first mission over Europe. P-61s also shot down some of the V1 flying bombs launched against Antwerp later in the year.

Massive for a fighter, the Widow was surprisingly agile due to its spoiler aileron control and near full-span, double-slotted flaps. Even in daylight dogfights, the P-61 could defeat most aircraft it went up against, including nimble Japanese fighters. It was also an outstanding ground support aircraft, able to carry 6,400 pounds (2,905 kg) of bombs externally, performing quite a bit of night attack. The basic airframe was modified into a more streamlined two-seat reconnaissance platform as the F-15 Reporter, which entered service in 1946.

Specification (P-61B)

Powerplants: twin Pratt & Whitney R-2800-65 Double Wasp 2,000 hp 18-cylinder radials.

Dimensions: length 15.12 m (49 ft 7 in); height 4.47 m (14 ft 8 in); wing span 20.12 m (66 ft).

Weights: empty 10,896 kg (24,000 lb); operational 17,252 kg (38,000 lb).

Performance: maximum speed 589 km/h (366 mph); service ceiling 10,065 m (33,000 ft); range 4,505 km (2,800 miles).

Armament: four 20 mm cannon, four .50 calibre machine guns in remote controlled top turret; external bomb load 2,905 kg (6,400 lb).

Republic P-47 Thunderbolt

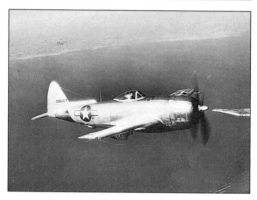

The P-47N was the final production version, the bubble canopy provided better visibility

Developed from a long line of Seversky/Republic fighters, the P-47 was initially designed as a standard weight airframe with an inline Allison engine. A redesign resulted in the hefty XP-47B, which flew for the first time in May 1941. Though initially denigrated for its large size, the P-47 quickly established itself as one of the finest Allied fighters over Europe, with high top speed and rugged construction which brought many home who would otherwise have gone down, and an effortlessness on the controls which surprised even the most critical of pilots. In a dogfight it could outroll almost anything and it seemed that a great deal of enemy fire had to be pumped into it to bring it down.

Early 'razorback' models later gave way to bubble canopy versions with improved visibility for the pilot. The final version of the Thunderbolt, the P-47N, had the range so many of its predecessors lacked. Sent to the Pacific, it could stay with the B-29s all the way to Japan and back or go on long-range strike missions to just about any target in the dwindling Japanese Empire. The massive Thunderbolt was America's heaviest single engine fighter, weighing in at over seven tons basic combat weight. Pilots who flew severely shot up Thunderbolts home swore they would never fly any other aircraft in combat.

Specification (P-47D)

Powerplant: Pratt & Whitney R-2800-59 Double Wasp 2,300 hp 18-cylinder radial.

Dimensions: length 11.01 m (36 ft 1.25 in); height 4.32 m (14 ft 2 in); wing span 12.43 m (40 ft 9.25 in).

Weights: empty 4,858 kg (10,700 lb); operational 8,807 kg (19,400 lb).

Performance: maximum speed 689 km/h (428 mph); service ceiling 12,810 m (42,000 ft); range 1,488 km (925 miles).

Armament: six .50 calibre machine guns; external load 1,135 kg (2,500 lb) of bombs, rockets or drop tanks.

Vought F4U/FG/F3A Corsair

A Vought Corsair II, built by Brewster and operated by the Royal Navy

When the prototype XF4U first flew in May 1940 it was the second American fighter (after the XP-38) to top 400 mph (644 km/h). The enormous propeller resulted in the famous 'bent wing' for ground clearance. Corsair deliveries to the U.S. Navy began in October 1942 but carrier landing trials were disappointing. The pilot was seated far aft with poor visibility from the 'birdcage' canopy and the landing gear shock struts were too stiff, producing a vicious bounce upon touchdown. As a result, the fighter was assigned to land-based squadrons with Marine VMF-124 getting the first aircraft into action over Bougainville in February 1943. By August seven more Marine squadrons were flying the Corsair and a